U.S. Army Combat Pistol Training Handbook

U.S. Army Combat Pistol Training Handbook

Department of the Army

Skyhorse Publishing

All inquiries should be addressed to Skyhorse Publishing, 307 West 36th Street, 11th Floor, New York, NY 10018.

Skyhorse Publishing books may be purchased in bulk at special discounts for sales promotion, corporate gifts, fund-raising, or educational purposes. Special editions can also be created to specifications. For details, contact the Special Sales Department, Skyhorse Publishing, 307 West 36th Street, 11th Floor, New York, NY 10018 or info@skyhorsepublishing.com.

Visit our website at www.skyhorsepublishing.com.

10 9 8 7 6 5 4 3 2 1

Library of Congress Cataloging-in-Publication Data is available on file.

ISBN: 978-1-62087-737-1

Printed in China

Contents

Preface vii

CHAPTER 1 COMPONENTS AND FUNCTIONING 1

Section I. Description and Components 1
1-1. Description 2
1-2. Components 3
1-3. Ammunition 4

Section II. Maintenance 5
1-4. Clearing Procedures 5
1-5. General Disassemble 6
1-6. Inspection 6
1-7. Cleaning, Lubrication, and Preventive Maintenance 6
1-8. General Assembly 7
1-9. Function Check 8

Section III. Operation and Function 9
1-10. Operation 9
1-11. Loading 10
1-12. Unloading and Clearing 11
1-13. Cycle of Operation 11

Section IV. Performance Problems 12
1-14. Malfunctions 12
1-15. Immediate Action 13
1-16. Remedial Action 13

CHAPTER 2 PISTOL MARKSMANSHIP TRAINING 15

Section I. Basic Marksmanship 15
2-1. Grip 15
2-2. Aiming 19
2-3. Breath Control 20
2-4. Trigger Squeeze 21
2-5. Target Engagement 23
2-6. Positions 23

Section II. Combat Marksmanship 29
2-7. Techniques of Firing 29
2-8. Target Engagement 31
2-9. Traversing 31

2-10. Combat Reloading Techniques 37
2-11. Poor Visibility Firing 39
2-12. Chemical, Biological, Radiological, or Nuclear 40

Section III. Coaching and Training Aids 40
2-13. Coaching 40
2-14. Ball-and-Dummy Method 41
2-15. Calling the Shot 41
2-16. Slow-Fire Exercise 42
2-17. Air-Operated Pistol, .177 MM 42
2-18. Quick-Fire Target Training Device 43
2-19. Range Firing Courses 47

Section IV. Safety 48
2-20. Requirements 49
2-21. Before Firing 49
2-22. During Firing 50
2-23. After Firing 50
2-24. Instructional Practice and Record Qualification Firing 50

* APPENDIX A COMBAT PISTOL QUALIFICATION COURSE 53
* APPENDIX B ALTERNATE PISTOL QUALIFICATION COURSE 65
APPENDIX C TRAINING SCHEDULES 73

GLOSSARY 77
REFERENCES 79
INDEX 81

* DA FORM 88-R, Combat Pistol Qualification Course Scorecard
* DA FORM 5704-R, Alternate Pistol Qualification Course Scorecard

Preface

This publication applies to the Active Army, the Army National Guard (ARNG), the National Guard of the United States (ARNGUS), and the US Army Reserve (USAR). It provides guidance on the operation and marksmanship of the M9, 9-mm pistol and the M11, 9-mm pistol. It reflects current Army standards in weapons qualification. It is a guide for the instructor to develop training programs, plans, and lessons that meet the objectives of the US Army Marksmanship program for developing combat-effective marksmen. The Soldier develops confidence, knowledge, and skills by following the guidelines in this manual.

The proponent of this publication is the US Army Infantry School. Send comments and recommendations on DA Form 2028 directly to Commandant, U.S. Army Infantry School, ATTN: ATSH-ATD, Fort Benning, GA 31905, or by email to doctrine@benning.army.mil.

Unless this publication states otherwise, masculine nouns and pronouns refer to either gender.

U.S. Army Combat Pistol Training Handbook

Chapter 1

Components and Functioning

This chapter describes the M9 and M11 semiautomatic pistols, their maintenance requirements, and their operation and functioning.

SECTION I. DESCRIPTION AND COMPONENTS

The M9 (Figure 1-1) and M11 (Figure 1-2) pistols are 9-mm, semiautomatic, magazine-fed, recoil-operation, double-action weapons chambered for the 9-mm cartridge.

Figure 1-1. 9-mm pistol, M9.

Figure 1-2. 9-mm pistol, M11.

1-1. Description

Table 1-1 summarizes equipment data for both pistols.

	M9 PISTOL	M11 PISTOL
Caliber	9-mm NATO	9-mm NATO
System of Operation	Short recoil, semiautomatic	Short recoil, semiautomatic
Locking System	Oscillating block	Oscillating block
Length	217 mm (8.54 inches)	180 mm (7.08 inches)
Width	38 mm (1.5 inches)	37 mm (1.46 inches)
Height	140 mm (5.51 inches)	136 mm (5.35 inches)
Magazine Capacity	**15 Rounds**	**13 Rounds**
Weight with Empty Magazine	960 grams (2.1 pounds)	745 grams (26.1 oz.)
Weight with 15-Round Magazine	1,145 grams (2.6 pounds)	830 grams (29.1 oz.)
Barrel Length	125 mm (4.92 inches)	98 mm (3.86 inches)
Rifling	Right-hand, six-groove (pitch 250 mm [about 10 inches])	Right-hand, six-groove (pitch 250 mm [9.84 inches])
Muzzle Velocity	375 meters per second (1,230.3 feet per second)	375 meters per second (1,230.3 feet per second)
Muzzle Energy	569.5 Newton meters (430 foot pounds)	569.5 Newton meters (430 foot pounds)
Maximum Range	1,800 meters (1,962.2 yards)	1,800 meters (1,962.2 yards)
Maximum Effective Range	50 meters (54.7 yards)	50 meters (54.7 yards)
Front Sight	Blade, integral with slide	Blade, integral with slide
Rear Sight	Notched bar, dovetailed to slide	Notched bar, dovetailed to slide
Sighting Radius	158 mm (6.22 inches)	158 mm (6.22 inches)
Safety Features	Decocking/safety lever, firing pin block	Decocking/safety lever, firing pin block
Hammer (half-cocked notch)	Prevents accidental discharge	Prevents accidental discharge
Basic Load	45 rounds	45 rounds
Trigger Pull	Single-action: 5.50 pounds Double-action: 12.33 pounds	Single-Action: 4.40 pounds Double-Action: 12.12 pounds

Table 1-1. Equipment Data, M9 and M11 pistols.

NOTE: For additional information on technical aspects of the M9 pistol, see TM 9-1005-317-10. For additional information on technical aspects of the M11 pistol, see TM 9-1005-325-10.

WARNING

The half-cocked position catches the hammer and prevents it from firing if the hammer is released while manually cocking the weapon. It is not to be used as a safety position. The pistol will fire from the half-cocked position if the trigger is pulled.

1-2. Components

The major components of the M9 (Figure 1-3) and M11 (Figure 1-4) pistols are:

a. **Slide and Barrel Assembly:** Houses the firing pin, striker, and extractor. Cocks the hammer during recoil cycle.
b. **Recoil Spring and Recoil Spring Guide:** Absorbs recoil and returns the slide assembly to its forward position.
c. **Barrel and Locking Block Assembly:** Houses cartridge for firing, directs projectile, and locks barrel in position during firing.
d. **Receiver:** Serves as a support for all the major components. Houses action of the pistol through four major components. Controls functioning of the pistol.
e. **Magazine:** Holds cartridges in place for stripping and chambering.

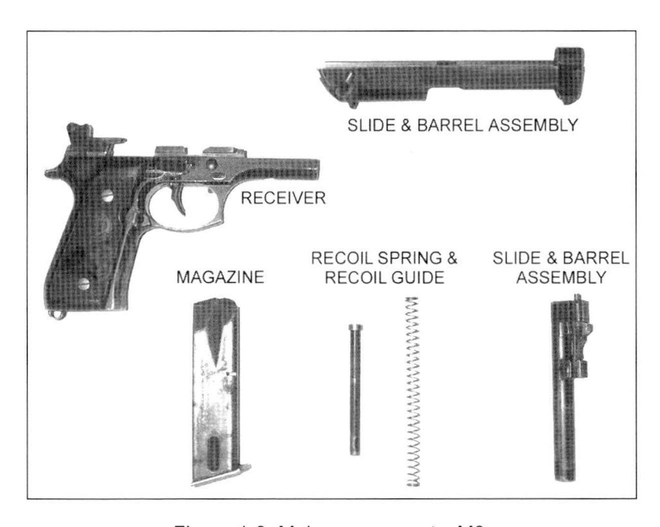

Figure 1-3. Major components, M9.

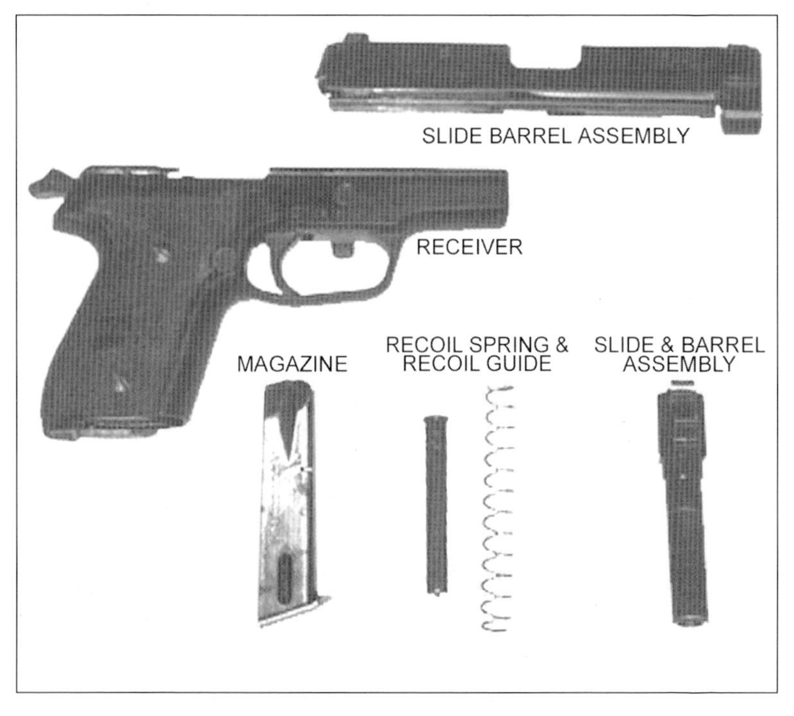

Figure 1-4. Major components, M11.

1-3. Ammunition

M9 and M11 pistols use several different types of 9-mm ammunition. Soldiers should use only authorized ammunition that is manufactured to US and NATO specifications.

a. **Type and Characteristics.** The specific type ammunition (Figure 1-5) and its characteristics are as follows:

(1) Cartridge, 9-mm ball, M882 with/without cannelure.

(2) Cartridge, 9-mm dummy, M917.

> **WARNING**
> **Do not fire heavily corroded or dented cartridges, cartridges with loose bullets, or any other rounds detected as defective through visual inspection.**

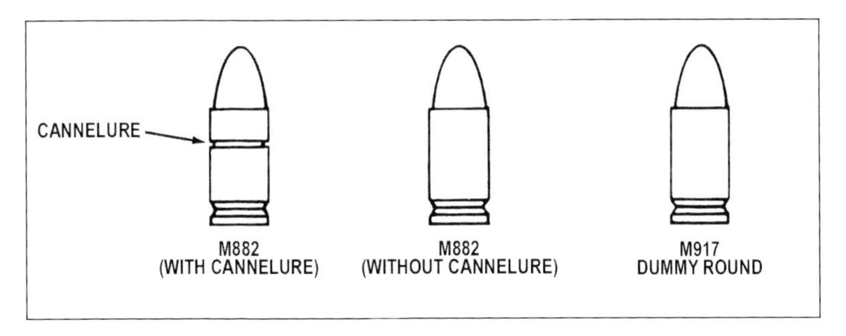

Figure 1-5. Ammunition.

b. **Care, Handling, and Preservation.**

(1) Protect ammunition from mud, sand, and water. If the ammunition gets wet or dirty, wipe it off at once with a clean dry cloth. Wipe off light corrosion as soon as it is discovered. Turn in heavily corroded cartridges.

(2) Do not expose ammunition to the direct rays of the sun. If the powder is hot, excessive pressure may develop when the pistol is fired.

(3) Do not oil or grease ammunition. Dust and other abrasives that collect on greasy ammunition may cause damage to the operating parts of the pistol. Oiled cartridges produce excessive chamber pressure.

SECTION II. MAINTENANCE

Maintenance procedures include clearing, disassembling, inspecting, cleaning, lubricating, assembling, and checking the functioning of the M9 or M11 pistol.

1-4. Clearing Procedures

The first step in maintenance is to clear the weapon. This applies in all situations, not just after firing. Soldiers must always assume the weapon is loaded. To clear the pistol, perform the following procedures.

a. Place the decocking/safety lever in the SAFE down position.
b. Hold the pistol in the raised pistol position.

c. Depress the magazine release button and remove the magazine from the pistol.
d. Pull the slide to the rear and remove any chambered round.
e. Push the slide stop up, locking the slide to the rear.
f. Look into the chamber to ensure that it is empty.

1-5. General Disassemble

To disassemble the pistol, perform the following procedures.

a. Depress the slide stop and let the slide go forward.
b. Hold the pistol in the right hand with the muzzle slightly raised.
c. Press the disassembly lever button with the forefinger.
d. Rotate the disassembly lever downward until it stops.
e. Pull the slide and barrel assembly forward and remove it from the receiver.
f. Carefully and lightly compress the recoil spring and spring guide. At the same time, lift up and remove them.
g. Separate the recoil spring from the spring guide.
h. Push in on the locking block plunger while pushing the barrel forward slightly.
i. Lift and remove the locking block and barrel assembly from the slide.

1-6. Inspection

Inspection begins with the pistol disassembled in its major components. Shiny surfaces do not mean the parts are unserviceable. Inspect all surfaces for visible damage, cracks, burrs, and chips.

1-7. Cleaning, Lubrication, and Preventive Maintenance

The M9 or M11 pistol should be disassembled into its major components and cleaned immediately after firing. All metal components and surfaces that have been exposed to powder fouling should be cleaned using CLP on a bore-cleaning patch. The same procedure is used to clean the receiver. After it has been cleaned and wiped dry, a thin coat of CLP is applied by rubbing with a cloth. This lubricates and preserves the exposed metal parts during all normal temperature ranges. When not in use, the pistol should be inspected weekly and cleaned and lubricated when necessary.

CAUTION
When using CLP, do not use any other type of cleaner. Never mix CLP with RBC or LSA.

a. Clear and disassemble the weapon.
b. Wipe or brush dirt, dust, and carbon buildup from the disassembled pistol.
c. Use CLP to help remove carbon buildup and stubborn dirt and grime.
d. Pay particular attention to the bolt face, guide rails on the receiver, grooves on the slide, and other hard-to-reach areas.

NOTE: Do not use mineral spirits, paint thinner, or dry cleaning solvent to clean the pistol. Use only issued lubricants and cleaners, such as CLP or LSA.

e. Clean the bore and chamber using CLP and fresh swabs.
f. Lubricate the pistol by covering all surfaces including the bore and chamber with a light coat of CLP. In extremely hot or cold weather, refer to the technical manual for lubricating procedures and materials.

1-8. General Assembly

To assemble the M9 or M11 pistol, simply reverse the procedures used to disassemble the pistol.

a. Grasp the slide with the bottom facing up.
b. With the other hand, grasp the barrel assembly with the locking block facing up.
c. Insert the muzzle into the forward end of the slide and, at the same time, lower the rear of the barrel assembly by aligning the extractor cutout with the extractor.

NOTE: The locking block will fall into the locked position in the slide.

d. Insert the recoil spring onto the recoil spring guide.

CAUTION
Maintain spring tension until the spring guide is fully seated in the cutaway on the locking block.

e. Insert the end of the recoil spring and the recoil spring guide into the recoil spring housing. At the same time, compress the recoil spring guide until it is fully seated on the locking block cutaway.

CAUTION
Do not pull the trigger while placing the slide on the receiver.

f. Ensure that the hammer is unlocked, the firing pin block is in the DOWN position, and the decocking/safety lever is in the SAFE position.
g. Grasp the slide and barrel assembly with the sights UP, and align the slide on the receiver assembly guide rails.
h. Push until the rear of the slide is a short distance beyond the rear of the receiver assembly and hold. At the same time, rotate the disassembly latch lever upward. A click indicates a positive lock.

1-9. Function Check

Always perform a function check after the pistol is reassembled to ensure it is working properly. To perform a function check:

a. Clear the pistol in accordance with the unloading procedures.
b. Depress the slide stop, letting the slide go forward.
c. Insert an empty magazine into the pistol.
d. Retract the slide fully and release it. The slide should lock to the rear.

e. Depress the magazine release button and remove the magazine.
f. Ensure the decocking/safety lever is in the SAFE position.
g. Depress the slide stop. When the slide goes forward, the hammer should fall to the forward position.
h. Squeeze and release the trigger. The firing pin block should move up and down and the hammer should not move.
i. Place the decocking/safety lever in the fire POSITION.
j. Squeeze the trigger to check double action. The hammer should cock and fall.
k. Squeeze the trigger again. Hold it to the rear. Manually retract and release the slide. Release the trigger. A click should be heard and the hammer should not fall.
l. Squeeze the trigger to check the single action. The hammer should fall.

SECTION III. OPERATION AND FUNCTION

This section provides detailed information on the functioning of M9 and M11 pistols.

1-10. Operation

With the weapon loaded and the hammer cocked, the shot is discharged by pulling the trigger.

a. Trigger movement is transmitted by the trigger bar, which draws the sear out of register with the full-cock hammer notch via the safety lever. With a slight timing lag, the safety lever also cams the safety lock upward to free the firing pin immediately before the hammer drops. The hammer forces the firing pin forward to strike and detonate the cartridge primer.
b. Blowback reaction generated by the exploding charge thrusts the locked barrel/slide system rearward against the recoil spring. After recoiling about 3 mm (1/8"), the barrel and slide unlock, allowing the barrel to tilt down into the locked position. The slide continues rearward until it abuts against the receiver stop.

c. During slide recoil, the hammer is cocked; the spent case is extracted and ejected as it strikes the ejector. In the initial recoil phase, the safety lever and safety lock separate, automatically rendering the firing pin safety lock effective again. As recoil continues, the slide depresses the trigger bar, disconnecting it from the safety lever. Sear spring pressure returns the sear and safety lever to their initial positions.
d. After contacting the receiver stop, the slide is thrust forward by the compressed recoil spring, stripping a round from the magazine and chambering it on the way. Just before reaching the forward end position, the slide again locks up with the barrel. The complete system is then thrust fully into the forward battery position by recoil spring pressure. Releasing the trigger allows the trigger bar and safety lever to re-engage.
e. The weapon is now cocked and ready to fire. After firing the last shot, the slide is locked in the rearmost position by the slide catch lever. This catch is actuated positively by the magazine follower, which is raised by magazine spring pressure.

1-11. Loading

To load the pistol–

- Hold the pistol in the raised pistol position.
- Insert the magazine into the pistol.
- Pull the slide to the rear and release the slide to chamber a round.
- Push the decocking/safety lever to the SAFE position.

a. Always make sure the muzzle is pointing in a safe direction, with the finger off the trigger.
b. Never attempt to load or unload any firearm inside a vehicle, building, or other confined space (except a properly constructed shooting range or bullet trap). Enclosed areas frequently offer no completely safe direction in which to point the firearm; if an accidental discharge occurs, there is great risk of injury or property damage.
c. Before loading, always clean excess grease and oil from the bore and chamber, and ensure that no obstruction is

in the barrel. Any foreign matter in the barrel could result in a bulged or burst barrel or other damage to the firearm and could cause serious injury to the shooter or to others.

1-12. Unloading and Clearing

To unload and clear the pistol–

- Hold the pistol in the raised pistol position.
- Depress the magazine release button and remove the magazine.
- Pull the slide to the rear and lock it in its rearward position by pushing up on the slide stop.
- Point the pistol skyward and look into the chamber to ensure it is clear.
- Let the slide go forward and pull the trigger to release the spring tension.

a. Perform this task in an area designated for this process.
b. Keep your finger off the trigger, and always make sure the muzzle is pointed in a safe direction.
c. Remember to clear the chamber after removing the magazine.
d. Never assume that a pistol is unloaded until you have personally checked it both visually and physically.
e. After every shooting practice, make a final check to be certain the firearm is unloaded before leaving the range.

1-13. Cycle of Operation

Each time a cartridge is fired, the parts inside the weapon function in a given order. This is known as the functioning cycle or cycle of operation. The cycle of operation of the weapon is divided into eight steps: feeding, chambering, locking, firing, unlocking, extracting, ejecting, and cocking. The steps are listed in the order in which functioning occurs; however, more than one step may occur at the same time.

a. A magazine containing ammunition is placed in the receiver. The slide is pulled fully to the rear and released. As the slide moves forward, it strips the top round from the magazine and pushes it into the chamber. The hammer remains in the cocked position, and the weapon is ready to fire.

b. The weapon fires one round each time the trigger is pulled. Each time a cartridge is fired, the slide and barrel recoil or move a short distance locked together. This permits the bullet and expanding powder gases to escape from the muzzle before the unlocking is completed.
c. The barrel then unlocks from the slide and continues to the rear, extracting the cartridge case from the chamber and ejecting it from the weapon. During this rearward movement, the magazine feeds another cartridge, the recoil spring is compressed, and the hammer is cocked.
d. At the end of the rearward movement, the recoil spring expands, forcing the slide forward, locking the barrel and slide together. The weapon is ready to fire again. The same cycle of operation continues until the ammunition is expended.
e. As the last round is fired, the magazine spring exerts upward pressure on the magazine follower. The stop on the follower strikes the slide stop, forcing it into the recess on the bottom of the slide and locking the slide to the rear. This action indicates that the magazine is empty and aids in faster reloading.

SECTION IV. PERFORMANCE PROBLEMS

Possible performance problems of M9 and M11 pistols are sluggish operation and stoppages. This section discusses immediate and remedial action to correct such problems.

1-14. Malfunctions

The following malfunctions may occur to the M9 and M11 pistols. Take these corrective actions to correct any problems that may occur.

a. **Sluggish Operation.** Sluggish operation is usually due to excessive friction caused by carbon build up, lack of lubrication, or burred parts. Corrective action includes cleaning, lubricating, inspecting, and replacing parts as necessary.

b. **Stoppages.** A stoppage is an interruption in the cycle of operation caused by faulty action of the pistol or faulty ammunition. Types of stoppages are:

- Failure to feed.
- Failure to chamber.
- Failure to lock.
- Failure to fire.
- Failure to unlock.
- Failure to extract.
- Failure to eject.
- Failure to cock.

1-15. Immediate Action

Immediate action is the action taken to reduce a stoppage without looking for the cause. Immediate action is taken within 15 seconds of a stoppage.

a. Ensure the decocking/safety lever is in the FIRE position.
b. Squeeze the trigger again.
c. If the pistol does not fire, ensure that the magazine is fully seated, retract the slide to the rear, and release.
d. Squeeze the trigger.
e. If the pistol again does not fire, remove the magazine and retract the slide to eject the chambered cartridge. Insert a new magazine, retract the slide, and release to chamber another cartridge.
f. Squeeze the trigger.
g. If the pistol still does not fire, perform remedial action.

1-16. Remedial Action

Remedial action is the action taken to reduce a stoppage by looking for the cause.

a. Clear the pistol.
b. Inspect the pistol for the cause of the stoppage.
c. Correct the cause of the stoppage, load the pistol, and fire.
d. If the pistol again fails to fire, disassemble it for closer inspection, cleaning, and lubrication.

CHAPTER 2

PISTOL MARKSMANSHIP TRAINING

Marksmanship training is divided into two phases: preparatory marksmanship training and range firing. Each phase may be divided into separate instructional steps. All marksmanship training must be progressive. Combat marksmanship techniques should be practiced after the basics have been mastered.

SECTION I. BASIC MARKSMANSHIP

The main use of the pistol is to engage an enemy at close range with quick, accurate fire. Accurate shooting results from knowing and correctly applying the elements of marksmanship. The elements of combat pistol marksmanship are:

- Grip.
- Aiming.
- Breath control.
- Trigger squeeze.
- Target engagement.
- Positions.

2-1. Grip

A proper grip is one of the most important fundamentals of quick fire. The weapon must become an extension of the hand and arm; it should replace the finger in pointing at an object. The firer must apply a firm, uniform grip to the weapon.

a. **One-Hand Grip**. Hold the weapon in the nonfiring hand; form a V with the thumb and forefinger of the strong hand (firing hand). Place the weapon in the V with the front and rear sights in line with the firing arm. Wrap the lower three fingers around the pistol grip, putting equal pressure with all three fingers to the rear. Allow the thumb of the firing hand to rest alongside the weapon without pressure (Figure 2-1). Grip the weapon tightly until the hand begins to tremble; relax until the trembling stops. At this point, the necessary pressure for a proper grip has been applied. Place the trigger finger on

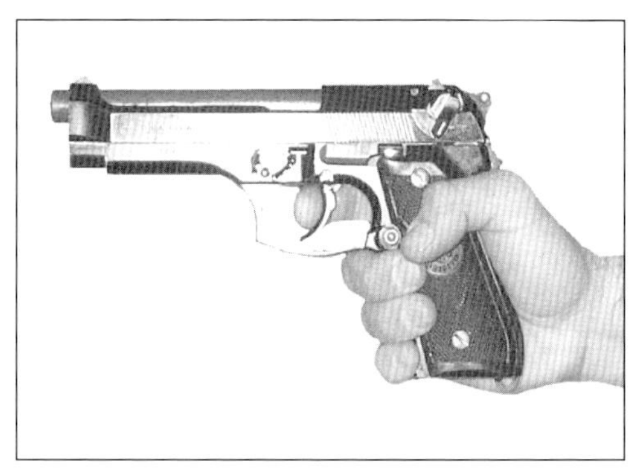

Figure 2-1. One-hand grip.

the trigger between the tip and second joint so that it can be squeezed to the rear. The trigger finger must work independently of the remaining fingers.

NOTE: If any of the three fingers on the grip are relaxed, the grip must be reapplied.

b. **Two-Hand Grip**. The two-hand grip allows the firer to steady the firing hand and provide maximum support during firing. The nonfiring hand becomes a support mechanism for the firing hand by wrapping the fingers of the nonfiring hand around the firing hand. Two-hand grips are recommended for all pistol firing.

WARNING
Do not place the nonfiring thumb in the rear of the weapon. The recoil upon firing could result in personal injury.

(1) ***Fist Grip***. Grip the weapon as with the one-hand grip. Firmly close the fingers of the nonfiring hand over the fingers of the firing hand, ensuring that the index finger from the nonfiring hand is between the middle finger of the firing hand and the trigger guard. Place the nonfiring thumb alongside the firing thumb (Figure 2-2).

NOTE: Depending upon the individual firer, he may chose to place the index finger of his nonfiring hand on the front of the trigger guard since M9 and M11 pistols have a recurved trigger guard designed for this purpose.

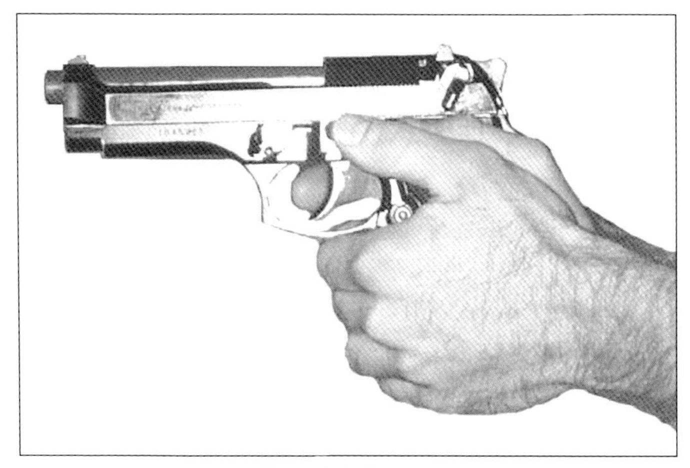

Figure 2-2. Fist grip.

(2) ***Palm-Supported Grip***. This grip is commonly called the cup and saucer grip. Grip the firing hand as with the one-hand grip. Place the nonfiring hand under the firing hand, wrapping the nonfiring fingers around the back of the firing hand. Place the nonfiring thumb over the middle finger of the firing hand (Figure 2-3).

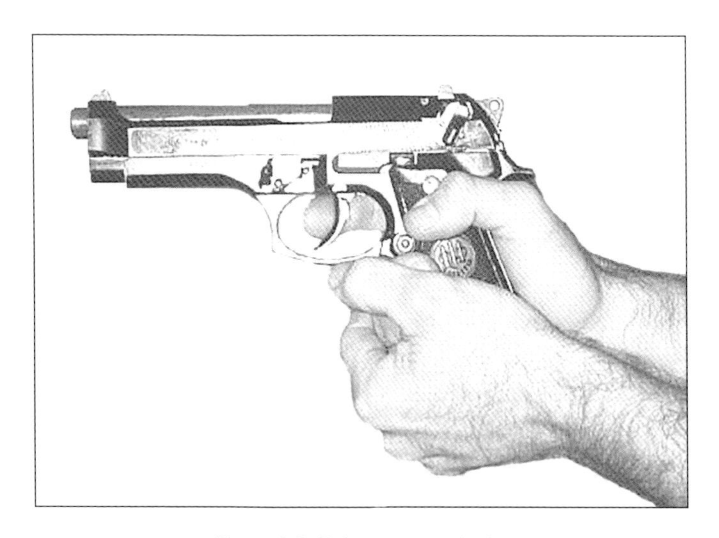

Figure 2-3. Palm-supported grip.

(3) ***Weaver Grip.*** Apply this grip the same as the fist grip. The only exception is that the nonfiring thumb is wrapped over the firing thumb (Figure 2-4).

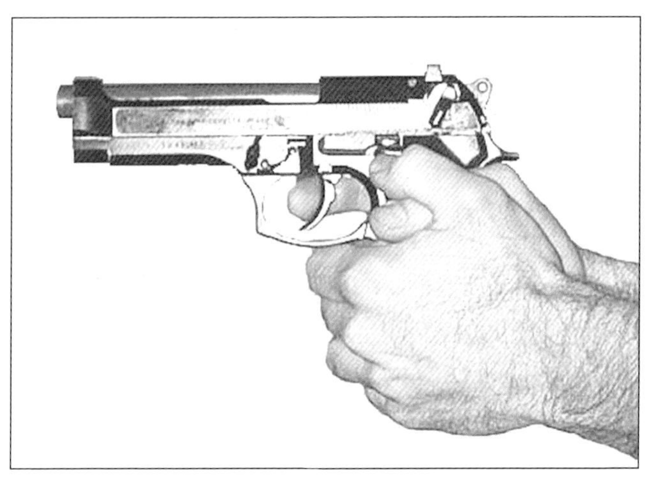

Figure 2-4. Weaver grip.

c. **Isometric Tension**. The firer raises his arms to a firing position and applies isometric tension. This is commonly known as the push-pull method for maintaining weapon stability. Isometric tension is when the firer applies forward pressure with the firing hand and pulls rearward with the nonfiring hand with equal pressure. This creates an isometric force but never so much to cause the firer to tremble. This steadies the weapon and reduces barrel rise from recoil. The supporting arm is bent with the elbow pulled downward. The firing arm is fully extended with the elbow and wrist locked. The firer must experiment to find the right amount of isometric tension to apply.

NOTE: The firing hand should exert the same pressure as the nonfiring hand. If it does not, a missed target could result.

d. **Natural Point of Aim**. The firer should check his grip for use of his natural point of aim. He grips the weapon and sights properly on a distant target. While maintaining his grip and stance, he closes his eyes for three to five seconds. He then opens his eyes and checks for proper sight picture. If the point of aim is disturbed, the firer adjusts his stance to compensate. If the sight alignment is disturbed, the firer adjusts his grip to compensate by removing the weapon from his hand and reapplying the grip. The firer repeats this process until the sight alignment and sight placement remain almost the same when he opens his eyes. With sufficient practice, this enables the firer to determine and use his natural point of aim, which is the most relaxed position for holding and firing the weapon.

2-2. Aiming

Aiming is sight alignment and sight placement (Figure 2-5).

a. Sight alignment is the centering of the front blade in the rear sight notch. The top of the front sight is level with the top of the rear sight and is in correct alignment with the eye. For correct sight alignment, the firer must center the front sight in the rear sight. He raises or lowers the top of the front sight so it is level with the top of the rear sight. Sight alignment is essential for accuracy because of the short sight radius of the pistol. For example, if a 1/10-inch error is made in aligning the front sight in the rear sight, the firer's bullet will miss the point of aim by about 15 inches at a range of 25 meters. The 1/10-inch error in sight alignment magnifies as the range increases–at 25 meters, it is magnified 150 times.

b. Sight placement is the positioning of the weapon's sights in relation to the target as seen by the firer when he aims the weapon (Figure 2-5). A correct sight picture consists of correct sight alignment with the front sight placed center mass of the target. The eye can focus on only one object at a time at different distances. Therefore, the last focus of the eye is always on the front sight. When the front sight is seen clearly, the rear sight and target will appear hazy. The firer can maintain correct sight alignment only through focusing on the front sight. His bullet will hit the target even if the sight picture is partly off center but still remains on the target. Therefore, sight alignment is more important than sight placement. Since it is impossible to hold the weapon completely still, the firer must apply trigger squeeze and maintain correct sight alignment while the weapon is moving in and around the center of the target. This natural movement of the weapon is referred to as wobble area. The firer must strive to control the limits of the wobble area through proper grip, breath control, trigger squeeze, and positioning.

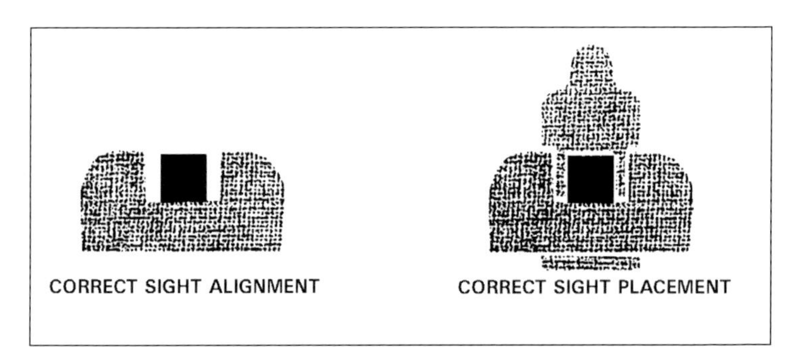

Figure 2-5. Correct sight alignment and sight picture.

c. Focusing on the front sight while applying proper trigger squeeze will help the firer resist the urge to jerk the trigger and anticipate the moment the weapon will fire. Mastery of trigger squeeze and sight alignment requires practice. Trainers should use concurrent training stations or have fire ranges to enhance proficiency of marksmanship skills.

2-3. Breath Control

To attain accuracy, the firer must learn to hold his breath properly at any time during the breathing cycle. This must be done while aiming and squeezing the trigger. While the procedure is simple, it requires explanation, demonstration, and supervised practice. To hold his breath properly, the firer takes a breath, lets it out, then inhales normally, lets a little out until comfortable, holds, and then fires. It is difficult to maintain a steady position keeping the front sight at a precise aiming point while breathing. Therefore, the firer should be taught to inhale, then exhale normally, and hold his breath at the moment of the natural respiratory pause (Figure 2-6). (Breath control, firing at a single target.) The shot must then be fired before he feels any discomfort from not breathing. When multiple targets are presented, the firer must learn to hold his breath at any part of the breathing cycle (Figure 2-7). Breath control must be practiced during dry-fire exercises until it becomes a natural part of the firing process.

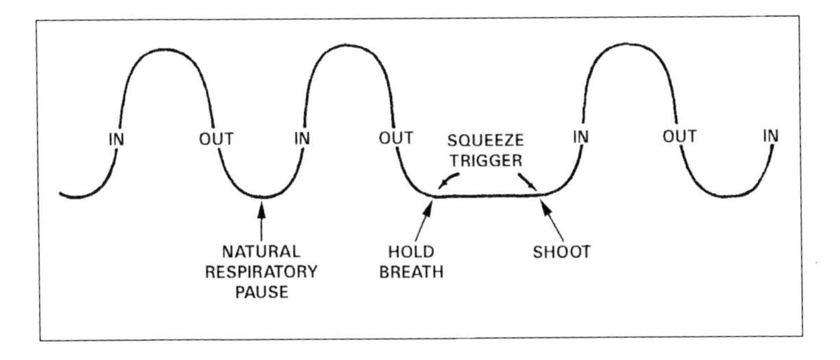

Figure 2-6. Breath control, firing at a single target.

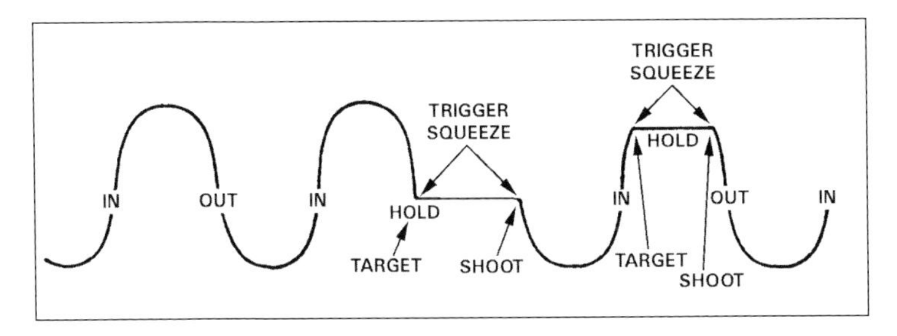

Figure 2-7. Breath control, firing at timed or multiple targets.

2-4. Trigger Squeeze

Improper trigger squeeze causes more misses than any other step of preparatory marksmanship. Poor shooting is caused by the aim being disturbed before the bullet leaves the barrel of the weapon. This is usually the result of the firer jerking the trigger or flinching. A slight off-center pressure of the trigger finger on the trigger can cause the weapon to move and disturb the firer's sight alignment. Flinching is an automatic human reflex caused by anticipating the recoil of the weapon. Jerking is an effort to fire the weapon at the precise time the sights align with the target. For more on problems in target engagement, see paragraph 2-5.

a. Trigger squeeze is the independent movement of the trigger finger in applying increasing pressure on the trigger straight to the rear, without disturbing the sight alignment until the weapon fires. The trigger slack, or free play, is taken up first, and the squeeze is continued steadily until the hammer falls. If the trigger is squeezed properly, the firer will not know exactly when the hammer will fall; thus, he will not tend to flinch or heel, resulting in a bad shot. Novice firers must be trained to overcome the urge to anticipate recoil. Proper application of the fundamentals will lower this tendency.

b. To apply correct trigger squeeze, the trigger finger should contact the trigger between the tip of the finger and the second joint (without touching the weapon anywhere else). Where contact is made depends on the length of the firer's trigger finger. If pressure from the trigger finger is applied to the right side of the trigger or weapon, the strike of the bullet will be to the left. This is due to the normal hinge action of the fingers. When the fingers on the right hand are closed, as in gripping, they hinge or pivot to the left, thereby applying pressure to the left (with left-handed firers, this action is to the right). The firer must not apply pressure left or right but should increase finger pressure straight to the rear. Only the trigger finger should perform this action. Dry-fire training improves a firer's ability to move the trigger finger straight to the rear without cramping or increasing pressure on the hand grip.

c. Follow-through is the continued effort of the firer to maintain sight alignment before, during, and after the round has fired. The firer must continue the rearward movement of the finger even after the round has been fired. Releasing the trigger too soon after the round has been fired results in an uncontrolled shot, causing a missed target.

(1) The firer who is a good shot holds the sights of the weapon as nearly on the target center as possible and continues to squeeze the trigger with increasing pressure until the weapon fires.

(2) The soldier who is a bad shot tries to "catch his target" as his sight alignment moves past the target and fires the weapon at that instant. This is called ambushing, which causes trigger jerk.

NOTE: The trigger squeeze of the pistol, when fired in the single-action mode, is 5.50 pounds; when fired in double-action mode, it is 12.33 pounds. The firer must be aware of the mode in which he is firing. He must also practice squeezing the trigger in each mode to develop expertise in both single-action and double-action target engagements.

2-5. Target Engagement

To engage a single target, the firer applies the method discussed in paragraph 2-4. When engaging multiple targets in combat, he engages the closest and most dangerous multiple target first and fires at it with two rounds. This is called controlled pairs. The firer then traverses and acquires the next target, aligns the sights in the center of mass, focuses on the front sight, applies trigger squeeze, and fires. He ensures his firing arm elbow and wrist are locked during all engagements. If he has missed the first target and has fired upon the second target, he shifts back to the first and engages it. Some problems in target engagement are as follows:

a. **Recoil Anticipation**. When a soldier first learns to shoot, he may begin to anticipate recoil. This reaction may cause him to tighten his muscles during or just before the hammer falls. He may fight the recoil by pushing the weapon downward in anticipating or reacting to its firing. In either case, the rounds will not hit the point of aim. A good method to show the firer that he is anticipating the recoil is the ball-and-dummy method (see paragraph 2-14).

b. **Trigger Jerk**. Trigger jerk occurs when the soldier sees that he has acquired a good sight picture at center mass and "snaps" off a round before the good sight picture is lost. This may become a problem, especially when the soldier is learning to use a flash sight picture (see paragraph 2-7b).

c. **Heeling**. Heeling is caused by a firer tightening the large muscle in the heel of the hand to keep from jerking the trigger. A firer who has had problems with jerking the trigger tries to correct the fault by tightening the bottom of the hand, which results in a heeled shot. Heeling causes the strike of the bullet to hit high on the firing hand side of the target. The firer can correct shooting errors by knowing and applying correct trigger squeeze.

2-6. Positions

The qualification course is fired from a standing, kneeling, or crouch position. During qualification and combat firing, soldiers must practice all of the firing positions described below so they become natural movements. Though these positions seem natural, practice sessions must be conducted to ensure the habitual attainment of correct firing positions. Practice in assuming correct firing positions ensures that soldiers can quickly assume these positions without a conscious effort. Pistol marksmanship requires a soldier to rapidly apply all the fundamentals at dangerously close targets while under stress. Assuming a proper position to allow for a steady aim is critical to survival.

NOTE: During combat, there may not be time for a soldier to assume a position that will allow him to establish his natural point of aim. Firing from a covered position may require the soldier to adapt his shooting stance to available cover.

a. **Pistol-Ready Position**. In the pistol-ready position, hold the weapon in the one-hand grip. Hold the upper arm close to the body and the forearm at about a 45-degree angle.

Figure 2-8. Pistol-ready position.

Point the weapon toward target center as you move forward (Figure 2-8).

b. **Standing Position without Support**. Face the target (Figure 2-9). Place feet a comfortable distance apart, about shoulder width. Extend the firing arm and attain a two-hand grip. The wrist and elbow of the firing arm are locked and pointed toward target center. Keep the body straight with the shoulders slightly forward of the buttocks.

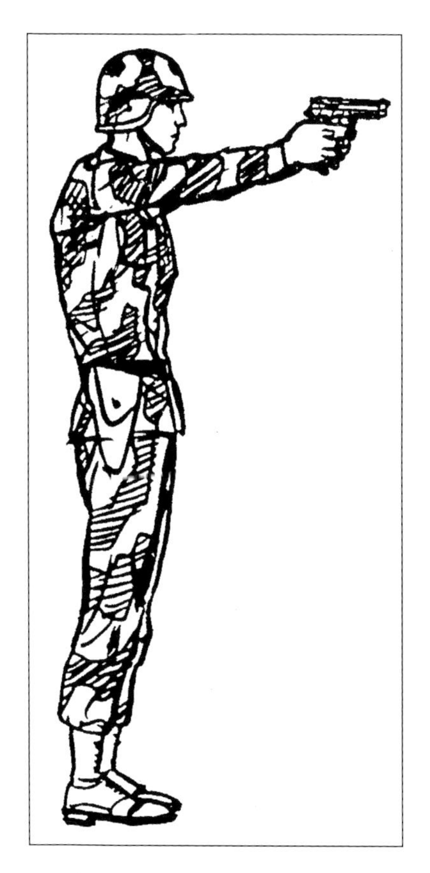

Figure 2-9. Standing position without support.

c. **Kneeling Position**. In the kneeling position, ground only your firing-side knee as the main support (Figure 2-10). Vertically place your firing-side foot, used as the main support,

Figure 2-10. Kneeling position.

under your buttocks. Rest your body weight on the heel and toes. Rest your nonfiring arm just above the elbow on the knee not used as the main body support. Use the two-handed grip for firing. Extend the firing arm, and lock the firing-arm elbow and wrist to ensure solid arm control.

d. **Crouch Position**. Use the crouch position when surprise targets are engaged at close range (Figure 2-11). Place the

Figure 2-11. Crouch position.

body in a forward crouch (boxer's stance) with the knees bent slightly and trunk bent forward from the hips to give faster recovery from recoil. Place the feet naturally in a position that allows another step toward the target. Extend the weapon straight toward the target, and lock the wrist and elbow of the firing arm. It is important to consistently train with this position, since the body will automatically crouch under conditions of stress such as combat. It is also a faster position from which to change direction of fire.

e. **Prone Position**. Lie flat on the ground, facing the target (Figure 2-12). Extend your arms in front with the firing arm locked. (Your arms may have to be slightly unlocked for firing at high targets.) Rest the butt of the weapon on the ground for single, well-aimed shots. Wrap the fingers of the nonfiring hand around the fingers of the firing hand. Face forward. Keep your head down between your arms and behind the weapon as much as possible.

Figure 2-12. Prone position.

f. **Standing Position with Support**. Use available cover for support–for example, a tree or wall to stand behind (Figure 2-13). Stand behind a barricade with the firing side on line with the edge of the barricade. Place the knuckles of the nonfiring fist at eye level against the edge of the barricade. Lock the elbow and wrist of the firing arm. Move the foot on the nonfiring side forward until the toe of the boot touches the bottom of the barricade.

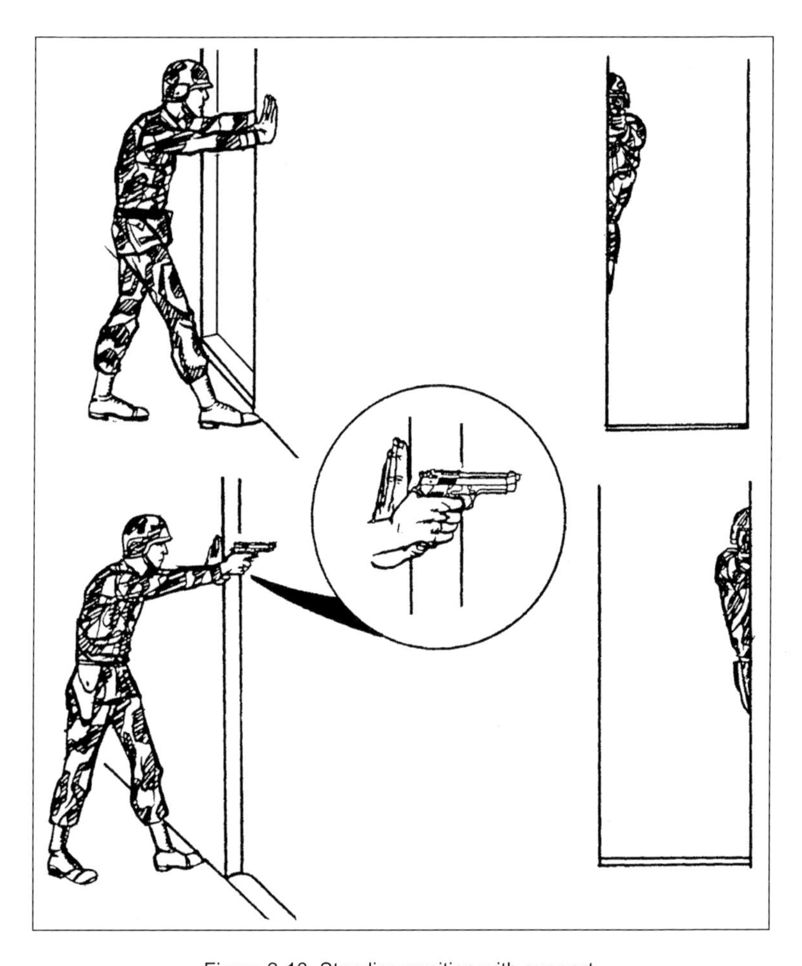

Figure 2-13. Standing position with support.

g. **Kneeling Supported Position**. Use available cover for support–for example, use a low wall, rocks, or vehicle (Figure 2-14). Place your firing-side knee on the ground. Bend the other knee and place the foot (nonfiring side) flat on the ground, pointing toward the target. Extend arms alongside and brace them against available cover. Lock the wrist and elbow of your firing arm. Place the nonfiring hand around the fist to support the firing arm. Rest the nonfiring arm just above the elbow on the nonfiring-side knee.

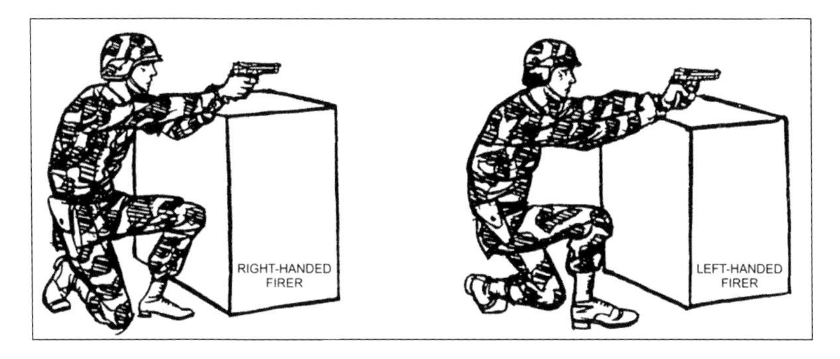

Figure 2-14. Kneeling supported.

SECTION II. COMBAT MARKSMANSHIP

After a soldier becomes proficient in the fundamentals of marksmanship, he progresses to advanced techniques of combat marksmanship. The main use of the pistol is to engage the enemy at close range with quick, accurate fire. In shooting encounters, it is not the first round fired that wins the engagement, but the first accurately fired round. The soldier should use his sights when engaging the enemy unless this would place the weapon within arm's reach of the enemy.

2-7. Techniques of Firing

Firing techniques include the use of hand-and-eye coordination, flash sight picture, quick-fire point shooting, and quick-fire sighting.

a. **Hand-and-Eye Coordination**. Hand-and-eye coordination is not a natural, instinctive ability for all soldiers. It is usually a learned skill obtained by practicing the use of a flash sight picture (see paragraph b below). The more a soldier practices raising the weapon to eye level and obtaining a flash sight picture, the more natural the relationship between soldier, sights, and target becomes. Eventually, proficiency elevates to a point so that the soldier can accurately engage targets in the dark. Each soldier must be aware of this trait and learn how to use it best. Poorly coordinated soldiers can achieve proficiency through close supervision from their trainers. Everyone has the ability to point at an object. Since pointing the forefinger at an object and extending the weapon toward a target are much the same, the combination of the two are natural. Making the soldier aware of this ability and teaching him how to apply it results in success when engaging enemy targets in combat.

(1) The eyes focus instinctively on the center of any object observed. After the object is sighted, the firer aligns his sights on the center of mass, focuses on the front sight, and applies proper trigger squeeze. Most crippling or killing hits result from maintaining the focus on the center of mass. The eyes must remain fixed on some part of the target throughout firing.

(2) When a soldier points, he instinctively points at the feature on the object on which his eyes are focused. An impulse from the brain causes the arm and hand to stop when the finger reaches the proper position. When the eyes are shifted to a new object or feature, the finger, hand, and arm also shift to this point. It is this inherent trait that can be used by the soldier to engage targets rapidly and accurately. This instinct is called hand-and-eye coordination.

b. **Flash Sight Picture**. Usually, when engaging an enemy at pistol range, the firer has little time to ensure a correct sight picture. The quick-kill (or natural point of aim) method does not always ensure a first-round hit. A compromise between a correct sight picture and the quick-kill method is known as a flash sight picture. As the soldier raises the weapon to eye level, his point of focus switches from the enemy to the front sight, ensuring that the front and rear sights are in proper alignment left and right, but not necessarily up and down. Pressure is applied to the trigger as the front sight is being acquired, and the hammer falls as the flash sight picture is confirmed. Initially, this method should be practiced slowly, with speed gained as proficiency increases.

c. **Quick-Fire Point Shooting**. This is for engaging an enemy at less than 5 yards and is also useful for night firing. Using a two-hand grip, the firer brings the weapon up close to the body until it reaches chin level. He then thrusts it forward until both arms are straight. The arms and body form a triangle, which can be aimed as a unit. In thrusting the weapon forward, the firer can imagine that there is a box between him and the enemy, and he is thrusting the weapon into the box. The trigger is smoothly squeezed to the rear as the elbows straighten.

d. **Quick-Fire Sighting**. This technique is for engaging an enemy at 5 to 10 yards away and only when there is no time available to get a full picture. The firing position is the same as for quick-fire point shooting. The sights are aligned left and right to save time, but not up and down. The firer must determine in practice what the sight picture will look like and where the front sight must be aimed to hit the enemy in the chest.

2-8. Target Engagement

In close combat, there is seldom time to precisely apply all of the fundamentals of marksmanship. When a soldier fires a round at the enemy, he often does not know if he hits his target. Therefore, two rounds should be fired at the target. This is called controlled pairs. If the enemy continues to attack, two more shots should be placed in the pelvic area to break the body's support structure, causing the enemy to fall.

2-9. Traversing

In close combat, the enemy may be attacking from all sides. The soldier may not have time to constantly change his position to adapt to new situations. The purpose of the crouching or kneeling 360-degree traverse is to fire in any direction without moving the feet.

a. **Crouching 360-Degree Traverse**. The following instructions are for a right-handed firer. The two-hand grip is used at all times except for over the right shoulder. The firer remains in the crouch position with feet almost parallel to each other. Turning will be natural on the balls of the feet.

(1) ***Over the Left Shoulder*** (Figure 2-15): The upper body is turned to the left, the weapon points to the left rear with the elbows of both arms bent. The left elbow is naturally bent more than the right elbow.

(2) ***Traversing to the Left*** (Figure 2-16): The upper body turns to the right, and the right firing arm straightens out. The left arm is slightly bent.

(3) ***Traversing to the Front*** (Figure 2-17): The upper body turns to the front as the left arm straightens out. Both arms are straight forward.

(4) ***Traversing to the Right*** (Figure 2-18): The upper body turns to the right as both elbows bend. The right elbow is naturally bent more than the left.

Figure 2-15. Traversing over the left shoulder.

Figure 2-16. Traversing to the left.

Figure 2-17. Traversing to the front.

Figure 2-18. Traversing to the right.

(5) ***Traversing to the Right Rear*** (Figure 2-19): The upper body continues to turn to the right until it reaches a

point where it cannot go further comfortably. Eventually the left hand must be released from the fist grip, and the firer will be shooting to the right rear with the right hand.

Figure 2-19. Traversing to the right rear.

b. **Kneeling 360-Degree Traverse**. The following instructions are for right-handed firers. The hands are in a two-hand grip at all times. The unsupported kneeling position is used. The rear foot must be positioned to the left of the front foot.

(1) ***Traversing to the Left Side*** (Figure 2-20): The upper body turns to a comfortable position toward the left. The weapon is aimed to the left. Both elbows are bent with the left elbow naturally bent more than the right elbow.

(2) ***Traversing to the Front*** (Figure 2-21): The upper body turns to the front, and a standard unsupported kneeling position is assumed. The right firing arm is straight, and the left elbow is slightly bent.

(3) ***Traversing to the Right Side*** (Figure 2-22): The upper body turns to the right as both arms straighten out.

(4) ***Traversing to the Rear*** (Figure 2-23): The upper body continues to turn to the right as the left knee is turned

to the right and placed on the ground. The right knee is lifted off the ground and becomes the forward knee. The right arm is straight, while the left arm is bent. The direction of the kneeling position has been reversed.

Figure 2-20. Traversing to the left, kneeling.

Figure 2-21. Traversing to the front, kneeling.

Figure 2-22. Traversing to the right, kneeling.

Figure 2-23. Traversing to the rear, kneeling.

(5) ***Traversing to the New Right Side*** (Figure 2-24): The upper body continues to the right. Both elbows are straight until the body reaches a point where it cannot go further comfortably. Eventually, the left hand must be released from the fist grip, and the firer is shooting to the right with the one-hand grip.

Figure 2-24. Traversing to the new right side, kneeling.

c. **Training Method**. This method can be trained and practiced anywhere and, with the firer simulating a two-hand grip, without a weapon. The firer should be familiar with firing in all five directions.

2-10. Combat Reloading Techniques

Overlooked as a problem for many years, reloading has resulted in many casualties due to soldiers' hands shaking or errors such as dropped magazines, magazines placed in the pistol backwards, or empty magazines placed back into the weapon. The stress state induced by a life-threatening situation causes soldiers to do things they would not otherwise do. Consistent, repeated training is needed to avoid such mistakes.

NOTE: These procedures should be used only in combat, not on firing ranges.

a. Develop a consistent method for carrying magazines in the ammunition pouches. All magazines should face down with the bullets facing forward and to the center of the body.

b. Know when to reload. When possible, count the number of rounds fired. However, it is possible to lose count in close combat. If this happens, there is a distinct difference in recoil of the pistol when the last round has been fired. Change magazines when two rounds may be left–one in the magazine and one in the chamber.

This prevents being caught with an empty weapon at a crucial time. Reloading is faster with a round in the chamber since time is not needed to release the slide.

c. Obtain a firm grip on the magazine. This precludes the magazine being dropped or difficulty in getting the magazine into the weapon. Ensure the knuckles of the hand are toward the body while gripping as much of the magazine as possible. Place the index finger high on the front of the magazine when withdrawing from the pouch. Use the index finger to guide the magazine into the magazine well.

d. Know which reloading procedure to use for the tactical situation. There are three systems of reloading: rapid, tactical, and one-handed. Rapid reloading is used when the soldier's life is in immediate danger and the reload must be accomplished quickly. Tactical reloading is used when there is more time and it is desirable to keep the replaced magazine because there are rounds still in it or it will be needed again. One-handed reloading is used when there is an arm injury.

(1) ***Rapid Reloading***.

(a) Place your hand on the next magazine in the ammunition pouch to ensure there is another magazine.

(b) Withdraw the magazine from the pouch while releasing the other magazine from the weapon. Let the replaced magazine drop to the ground.

(c) Insert the replacement magazine, guiding it into the magazine well with the index finger.

(d) Release the slide, if necessary.

(e) Pick up the dropped magazine if time allows. Place it in your pocket, not back into the ammunition pouch where it may become mixed with full magazines.

(2) ***Tactical Reloading***.

(a) Place your hand on the next magazine in the ammunition pouch to ensure there is a remaining magazine.

(b) Withdraw the magazine from the pouch.

(c) Drop the used magazine into the palm of the nonfiring hand, which is the same hand holding the replacement magazine.

(d) Insert the replacement magazine, guiding it into the magazine well with the index finger.

(e) Release the slide, if necessary.

(f) Place the used magazine into a pocket. Do not mix it with full magazines.

(3) ***One-Hand Reloading, Right Hand***.
(a) Push the magazine release button with the thumb.
(b) Place the safety ON with the thumb if the slide is forward.
(c) Place the weapon backwards into the holster.

NOTE: If placing the weapon in the holster backwards is a problem, place the weapon between the calf and thigh to hold the weapon.

(d) Insert the replacement magazine.
(e) Withdraw the weapon from the holster.
(f) Remove the safety with the thumb if the slide is forward, or push the slide release if the slide is back.
(4) ***One-Hand Reloading, Left Hand***.
(a) Push the magazine release button with the middle finger.
(b) Place the weapon backwards into the holster.

NOTE: If placing the weapon in the holster backwards is a problem, place the weapon between the calf and thigh to hold the weapon.

(c) Insert the replacement magazine.
(d) Remove the weapon from the holster.
(e) Remove the safety with the thumb if the slide is forward, or push the slide release lever with the middle finger if the slide is back.

2-11. Poor Visibility Firing

Poor visibility firing with any weapon is difficult since shadows can be misleading to the firer. This is mainly true during EENT and EMNT (a half hour before dark and a half hour before dawn). Even though the pistol is a short-range weapon, the hours of darkness and poor visibility further decrease its effect. To compensate, the firer must use the three principles of night vision:

a. **Dark Adaptation**. This process conditions the eyes to see during poor visibility conditions. The eyes usually need about 30 minutes to become 90 percent adapted in a totally darkened area.

b. **Off-Center Vision**. When looking at an object in daylight, a person looks directly at it. However, at night he would see the object only for a few seconds. To see an object in darkness, he must concentrate on it while looking 6 to 10 degrees away from it.

c. **Scanning**. This is the short, abrupt, irregular movement of the firer's eyes around an object or area every 4 to 10 seconds. With artificial illumination, the firer uses night-fire techniques to engage targets, since targets seem to shift without moving.

NOTE: For more detailed information on the three principles of night vision, see FM 21-75 (US Army Field Manual 21-75).

2-12. Chemical, Biological, Radiological, or Nuclear

When firing a pistol under CBRN conditions, the firer should use optical inserts, if applicable. Firing in MOPP levels 1 through 3 should not be a problem for the firer. Unlike with a rifle, the firer acquires a sight picture with a pistol the same with or without a protective mask. MOPP4 is the only level that might present a problem for a firer, because that level requires him to wear gloves. Gloves could force him to adjust for proper grip and trigger squeeze. Firers should practice firing in MOPP4 to become proficient in CBRN firing.

SECTION III. COACHING AND TRAINING AIDS

Throughout preparatory marksmanship training, the coach-and-pupil method of training should be used. The proficiency of a pupil depends on how well the coach performs his duties. This section provides detailed information on coaching techniques and training aids for pistol marksmanship.

2-13. Coaching

The coach assists the firer by correcting errors, ensuring he takes proper firing positions, and ensuring he observes all safety precautions. The criteria for selecting coaches are a command responsibility. Coaches must have more experience in pistol marksmanship than the student firer. Duties of the coach during instructional practice and record fire include the following:

a. Checking that the—

- Weapon is clear.

- Ammunition is clean.
- Magazines are clean and operational.

b. Observing the firer to see that he—

- Takes the correct firing position.
- Loads the weapon properly and only on command.
- Takes up the trigger slack correctly.
- Squeezes the trigger correctly.
- Calls the shot each time he fires, except during quick fire and rapid fire.
- Holds his breath correctly.
- When he does not fire for 5 or 6 seconds, lowers the weapon and rests his arm.

c. Having the firer breathe deeply several times to relax if he is tense.

2-14. Ball-and-Dummy Method

In this method, the coach loads the weapon for the firer. He may hand the firer a loaded weapon or an empty one. When firing the empty weapon, the firer observes that in anticipating recoil he is forcing the weapon downward as the hammer falls. Repetition of the ball-and-dummy method helps reduce recoil anticipation.

2-15. Calling the Shot

To call the shot is to state where the bullet should strike the target according to the sight picture at the instant the weapon fires, for example, "High," "a little low," "to the left," or "bull's eye." Another method of calling the shot is the clock system, for example, "three-ring hit at 8 o'clock" or "four-ring hit at 5 o'clock." Another method is to place a firing center beside the firer on the firing line. As soon as the shot is fired, the firer must place a finger on the target face or center where he expects the round to hit on the target. This method avoids guessing and computing for the firer. The immediate placing of the finger on the target face gives an accurate call. If the firer calls his shot incorrectly in range fire, he is failing to concentrate on sight alignment and trigger squeeze. Thus, as the weapon fires, he does not know what his sight picture is.

2-16. Slow-Fire Exercise

The slow-fire exercise is one of the most important exercises for both amateur and competitive marksmen. Coaches should ensure firers practice this exercise as much as possible. This is a dry-fire exercise.

a. To perform the slow-fire exercise, the firer assumes the standing position with the weapon pointed at the target. The firer should begin by using the two-hand grip, progressing to the one-hand grip as his skill increases. He takes in a normal breath and lets out part of it, locking the remainder in his lungs by closing his throat. He then relaxes, aims at the target, and takes the correct sight alignment and sight picture. He takes up the trigger slack and squeezes the trigger straight to the rear with steady, increasing pressure until the hammer falls, simulating firing.

b. If the firer does not cause the hammer to fall in 5 or 6 seconds, he should return to the pistol-ready position and rest his arm and hand. He then starts the procedure again. The action sequence that makes up this process can be summed up by the key word BRASS. It is a word the firer should think of each time he fires his weapon.

Breathe	Take a normal breath, let part of it out, and lock the remainder in the lungs by closing the throat.
Relax	Relax the body muscles.
Aim	Take correct sight alignment and sight picture, and focus the eye at the top of the front sight.
Slack	Take up the trigger slack.
Squeeze	Squeeze the trigger straight to the rear with steadily increasing pressure without disturbing sight alignment until the hammer falls.

c. Coaches should observe the front sight for erratic movements during the application of trigger squeeze. Proper application of trigger squeeze allows the hammer to fall without the front sight moving. A small bouncing movement of the front sight is acceptable. Firers should call the shot by the direction of movement of the front sight (high, low, left, or right).

2-17. Air-Operated Pistol, .177 MM

The air-operated pistol is used as a training device to teach the soldier the method of quick fire, to increase confidence in his ability, and to afford him more practice firing. A range can be set up almost anywhere with a minimum of effort and coordination, which is ideal for USAR and NG. If conducted on a standard range, live firing of pistols can be conducted along with the firing of the .177-mm air-operated pistol. Due to light recoil and little noise of the pistol, the soldier can concentrate on fundamentals. This helps build confidence because the soldier can hit a target faster and more accurately. The air-operated pistol should receive the same respect as any firearm. A thorough explanation of the weapon and a safety briefing are given to each soldier.

2-18. Quick-Fire Target Training Device

The QTTD (Figures 2-25 and 2-26) is used with the .177-mm air-operated pistol.

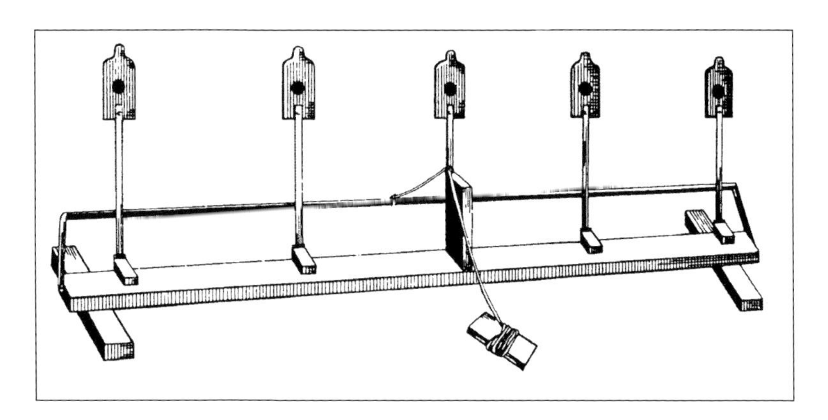

Figure 2-25. The quick-fire target training device.

a. **Phase I**. From 10 feet, five shots at a 20-foot miniature E-type silhouette. After firing each shot, the firer and coach discuss the results and make corrections.

b. **Phase II**. From 15 feet, five shots at a 20-foot miniature E-type silhouette. The same instructions apply to this exercise as for Phase I.

c. **Phase III**. From 20 feet, five shots at a 20-foot miniature E-type silhouette. The same instructions apply to this exercise as for Phases I and II.

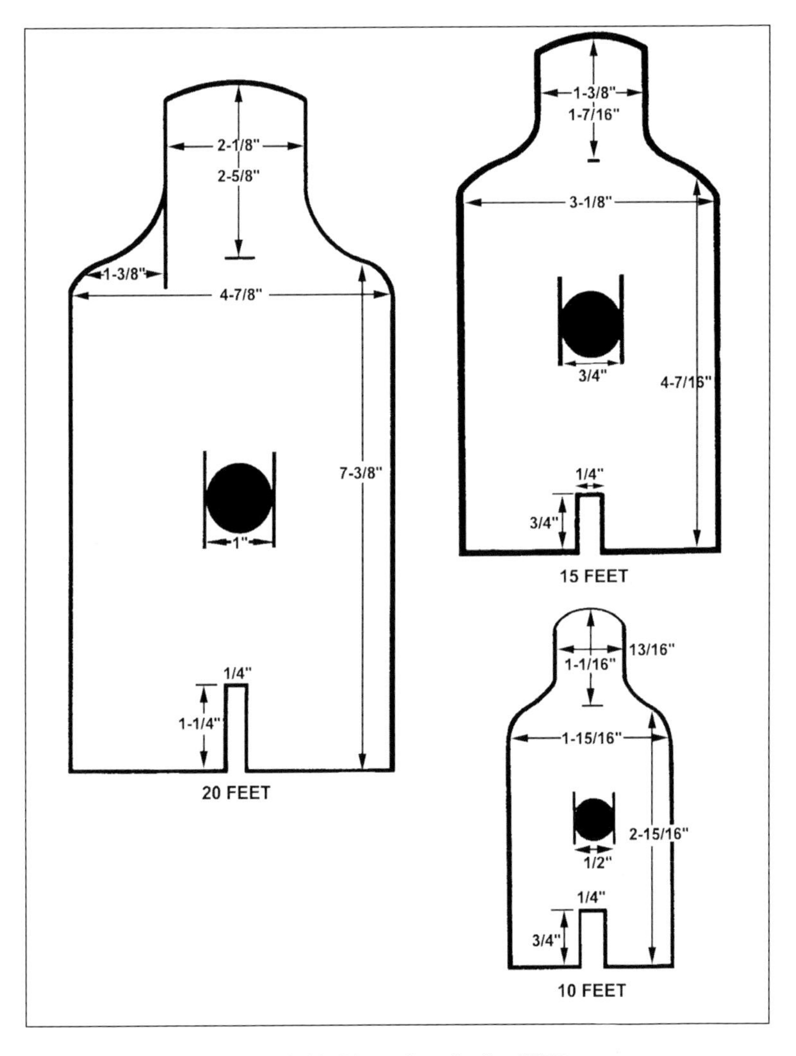

Figure 2-26. Dimensions for the QTTD.

d. **Phase IV**. From 15 feet, six shots at two 20-foot miniature E-type silhouettes. This exercise is conducted the same as the previous one, except that the firer is introduced to fire distribution. The targets on the QTTD are held in the up position so they cannot be knocked down when hit.

(1) The firer first engages the 20-foot miniature E-type silhouette on the extreme right of the QTTD (see Figure 2-27).

He then traverses between targets and engages the same type target on the extreme left of the QTTD. The firer again shifts back to reengage the first target. The procedure is used to teach the firer to instinctively return to the first target if he misses it with his first shot.

(2) The firer performs this exercise twice, firing three shots each time. Before firing the second time, the coach and firer should discuss the errors made during the first exercise.

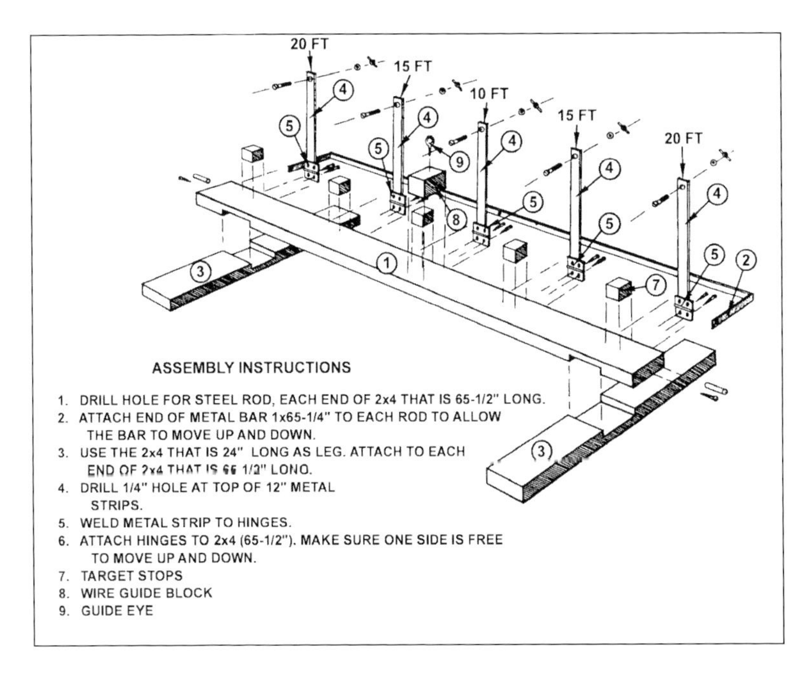

Figure 2-27. Miniature E-type silhouette for use with QTTD.

e. **Phase V**. Seven shots fired from 20, 15, and 10 feet at miniature E-type silhouettes.

(1) The firer starts this exercise 30 feet from the QTTD. The command MOVE OUT is given, and the firer steps out at a normal pace with the weapon held in the ready position. Upon the command FIRE (given at the 20-foot line), the firer assumes the crouch position and engages the 20-foot miniature E-type silhouette on the extreme right of the QTTD. He then traverses between targets,

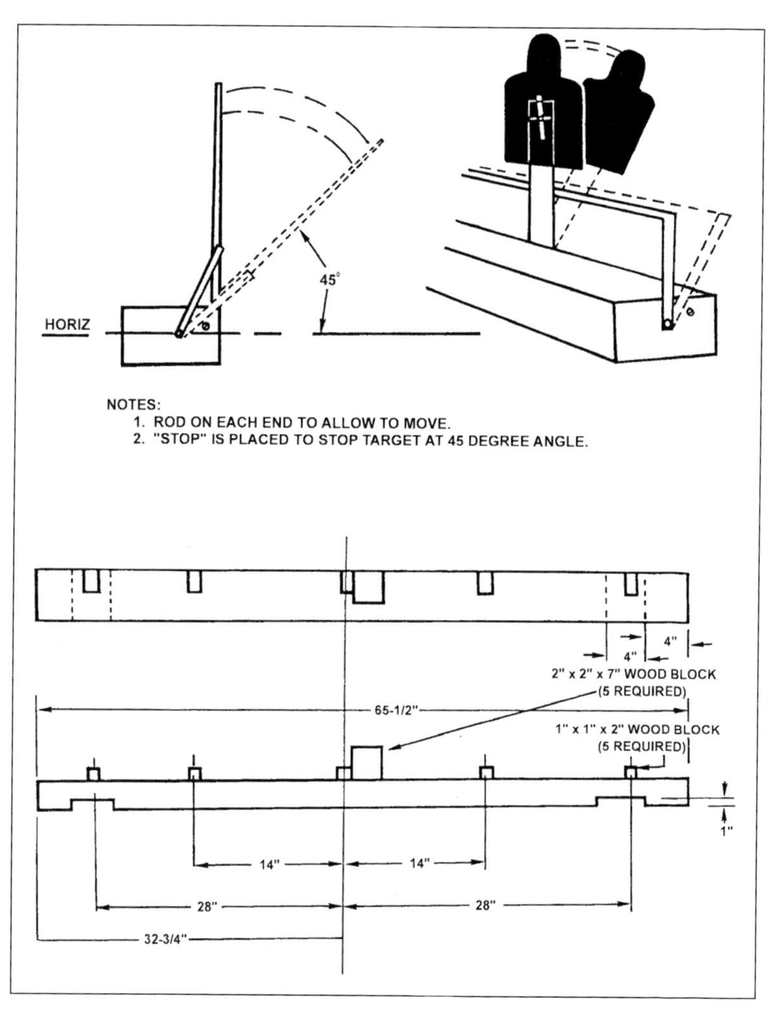

Figure 2-27. Miniature E-type silhouette for use with QTTD (continued).

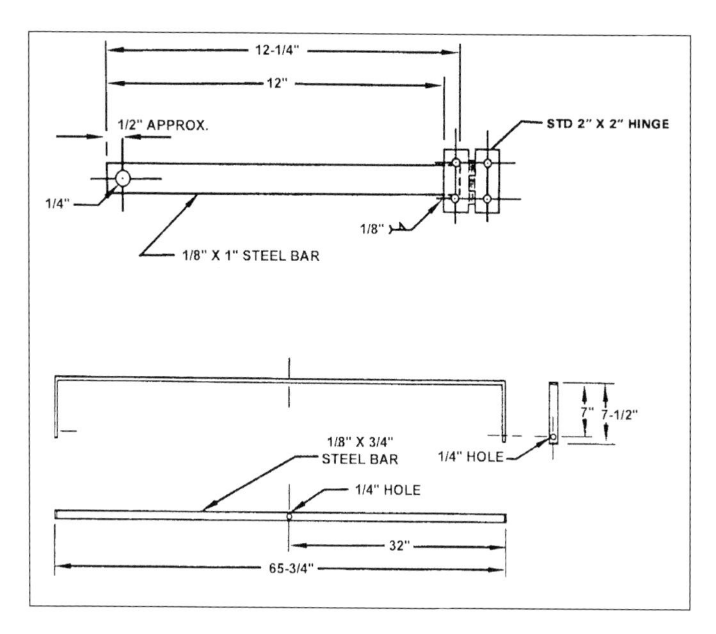

Figure 2-27. Miniature E-type silhouette for use with QTTD (continued).

engages the same type target on the extreme left of the QTTD, and shifts back to the first target. If the target is still up, he engages it. The firer then assumes the standing position and returns the weapon to the ready position. (Upon completion of each exercise, the coach makes corrections as the firer returns to the standing position.)

(2) On the command MOVE OUT, the firer again steps off at a normal pace. Upon the command FIRE (given at the 15-foot line), he engages the 15-foot targets on the QTTD. The same sequence of fire distribution is followed as with the previous exercise.

(3) During this exercise, the firer moves forward on command until he reaches the 10-foot line. At the command FIRE, the firer engages the 10-foot miniature E-type silhouette in the center of the QTTD.

2-19. Range Firing Courses

Range firing is conducted after the firers have satisfactorily completed preparatory marksmanship training. The range firing courses are:

a. **Instructional**. Instructional firing is practice firing on a range, using the assistance of a coach.

(1) All personnel authorized or required to fire the pistol receive 12 hours of preliminary instruction that includes the following:

- Disassembly and assembly.
- Loading, firing, unloading, and immediate action.
- Preparatory marksmanship.
- Care and cleaning.

(2) The tables fired for instructional practice are prescribed in the combat pistol qualification course in Appendix A. During the instructional firing, the CPQC is fired with a coach or instructor.

b. **Combat Pistol Qualification**. The CPQC stresses the fundamentals of quick fire. It is the final test of a soldier's proficiency and the basis for his marksmanship classification. After the soldier completes the instructional practice firing, he shoots the CPQC for record. Appendix A provides a detailed description of the CPQC tables, standards, and conduct of fire. TC 25-8 provides a picture of the course.

NOTE: The alternate pistol qualification course (APQC) can be used for sustainment/qualification if the CPQC is not available (see Appendix B).

c. **Military Police Firearms Qualification**. The military police firearms qualification course is a practical course of instruction for police firearms training (see FM 19-10).

SECTION IV. SAFETY

Safety must be observed during all marksmanship training. Listed below are the precautions for each phase of training. It is not intended to replace AR 385-63 or local range regulations. Range safety requirements vary according to the requirements of the course of fire. It is mandatory that the latest range safety directives and local range regulations be consulted to determine current safety requirements.

2-20. Requirements

The following requirements apply to all marksmanship training.

a. Display a red flag prominently on the range during all firing.
b. Soldiers must handle weapons carefully and never point them at anyone except the enemy in actual combat.
c. Always assume a weapon is loaded until it has been thoroughly examined and found to contain no ammunition.
d. Indicate firing limits with red and white striped poles visible to all firers.
e. Never place obstructions in the muzzle of any weapon about to be fired.
f. Keep weapons in a prescribed area with proper safeguards.
g. Refrain from smoking on the range near ammunition, explosives, or flammables.

2-21. Before Firing

The following requirements must be met before conducting marksmanship training.

a. Close and post guards at all prescribed roadblocks and barriers.
b. Ensure all weapons are clear of ammunition and obstructions, and all slides are locked to the rear.
c. Brief all firers on the firing limits of the range and firing lanes. Firers must keep their fires within prescribed limits.
d. Ensure all firers receive instructions on know how to load and unload the weapon and on safety features.
e. Brief all personnel on all safety aspects of fire and of the range pertaining to the conduct of the courses.
f. No one moves forward of the firing line without permission of the tower operator, safety officer, or OIC.
g. Weapons are loaded and unlocked only on command from the tower operator except during conduct of the courses requiring automatic magazine changes.
h. Weapons are not handled except on command from the tower operator.
i. Firers must keep their weapons pointed downrange when loading, preparing to fire, or firing.

2-22. During Firing

The following requirements apply during marksmanship training.

a. A firer does not move from his position until his weapon has been cleared by safety personnel and placed in its proper safety position. An exception is the assault phase.
b. During Table 5 of the CPQC, firers remain on line with other firers on their right or left.
c. Firers must fire only in their own firing lane and must not point the weapon into an adjacent lane, mainly during the assault phase.
d. Firers treat the air-operated pistol as a loaded weapon, observing the same safety precautions as with other weapons.
e. All personnel wear helmets during live-fire exercises.
f. Firers hold the weapon in the raised position except when preparing to fire. They then hold weapons in the ready position, pointed downrange.

2-23. After Firing

Safety personnel inspect all weapons to ensure they are clear. A check is conducted to determine if any brass or live ammunition is in the possession of the soldiers. Once cleared, pistols are secured with the slides locked to the rear.

2-24. Instructional Practice and Record Qualification Firing

During these phases of firing, safety personnel ensure that—

a. The firer understands the conduct of the exercise.
b. The firer has the required ammunition and understands the commands for loading and unloading.
c. The firer complies with all commands from the tower operator.
d. Firers maintain proper alignment with other firers while moving downrange.
e. Weapons are always pointed downrange.
f. Firers fire within the prescribed range limits.
g. Weapons are cleared after each phase of firing, and the tower-operator is aware of the clearance.

h. Malfunctions or failures to fire that are due to no fault of the firer are reported immediately. On command of the tower operator, the weapon is cleared and action is taken to allow the firer to continue with the exercise.

NOTE: For training and qualification standards, see Appendixes A through D.

APPENDIX A*

COMBAT PISTOL QUALIFICATION COURSE

This appendix explains the combat pistol qualification course. If it is unavailable, the alternate pistol qualification course (APQC) may be used to sustain training and to qualify firers.

The tower operator is completely responsible for and in charge of the range and the course. He controls absolutely all activities related to firing. The tower operator tells the scorers what to do when, for example, when to issue the preloaded magazines to firers. Only the tower operator may issue the order to fire. Scorers and firers must await the tower operator's orders.

A-1. Course Information

The CPQC (shown in TC 25-8) requires the Soldier to engage single and multiple targets at various ranges using the fundamentals of quick fire.

a. **Extra Rounds.** For each table of the CPQC, the firer is given extra rounds to reengage missed targets. Although only 30 targets will be exposed during the entire course, each firer will receive 40 rounds of ammunition. Hitting a target with an additional round during the exposure time is just as effective as hitting it with the first round. Consequently, the firer is not penalized for using or not using the extra ammunition. However, any unused ammunition must be turned in at the end of the table, and may not be used in any other table.

b. **Magazine Changes.** Only three magazine changes are required during this course: one change in Firing Table II, and two changes in Firing Table V. For safety, each of these two tables begins with a magazine loaded only with one round. The first target appears, and the firer engages it with that round. By the time another target appears eight seconds later, the firer must have reloaded and prepared to engage. He will receive no commands to reload. Failure to reload in time to engage the second target is scored as a miss. This teaches the Soldier to change magazines instinctively, quickly, and safely under pressure. In Table V, a second magazine change is commanded by the control tower.

c. **Double-Action Mode**. When firing the 9-mm pistol, the Soldier uses double-action to fire the first round in every table.

d. **Range to Target**. The range to exposed targets must not exceed 31 meters from the firer. Table A-1 shows target exposure times for each firing table.

NUMBER OF TARGETS	FIRING TABLE						
	I	II	III	IV	V	VI	VII
Single targets	3 Sec			2 Sec		10 Sec	
Multiple targets	5 Sec			4 Sec		20 Sec	

Table A-1. Target-exposure times.

A-2. Standards by Firing Table

The following qualification tables apply for day, night, and CBRN qualification. The standing firing position is used throughout the qualification.

NOTE: 1. The range OIC determines a common target sequence for all lanes. This keeps a firer from getting ahead of adjacent firers.
2. Target sequences vary in distance from the firer, starting with no more than two targets at 7 meters and the farthest targets at 31 meters.

a. **Table I—Day Standing**. For this table, the firer receives one magazine with seven rounds in it. Five targets (single) are exposed. The firer assumes the standing firing position at the firing line. He holds the weapon at the ready. The tower operator sets the target sequence.

b. **Table II—Day Standing**. For this table, the firer receives two magazines: one containing one round, and the other containing seven rounds. Six targets (four single and one set of two) are exposed. The firer takes the same position on the firing line as he did in Table I.

(1) ***First Magazine***. The firer loads the first magazine (containing one round). One target is exposed.

(2) ***Second Magazine***. After he fires the round in the first magazine, the firer must change magazines at once. He has eight seconds to load the second magazine (containing seven rounds) and prepare to fire before the next target is exposed. Once it appears, he must engage in the three seconds before it is lowered. Failure to do so is scored as a miss.

c. **Table III–Day Standing**. For this table, the firer receives one magazine containing seven rounds. Five targets (three single and one set of two) are exposed. The firer fires at each target, or set of targets, and rotates to the next firing point for that table.

d. **Table IV–Day Standing**. For this table, the firer receives one magazine containing five rounds. Four targets are exposed. The firer starts in the same position used in the previous tables. Four targets (two single targets and one set two more) are exposed to the firer.

e. **Table V–Day Moving Out**. For this table, the firer receives three magazines: one each with one, seven, and five rounds. Ten targets are exposed. The firer begins 10 meters behind the firing line, in the middle of the trail.

(1) The firer loads the first magazine (containing one round). He places the second magazine (containing seven rounds) in the magazine pouch closest to his firing hand. He places third magazine (containing five rounds) in the magazine pouch farthest from his firing hand.

(2) When the firer reaches the firing line, a single target is exposed. The firer has two seconds to hit it before it is lowered. He then has eight seconds to load the second magazine (containing seven rounds).

(3) At the end of eight seconds, another single target is exposed to the firer. If the firer has not loaded the second magazine in time to engage this target, this round is scored as a miss.

(4) When the tower operator is sure that the firing line has completed the magazine change, he commands MOVE

OUT. He then exposes two multiple targets, one after the other, at various ranges from the firer.

(5) After two sets of multiple targets are exposed, the Soldier is commanded to load the five-round magazine. After the command MOVE OUT is given, the remaining targets are presented to the firer in sequence. After the last targets are hit or lowered, the firer clears the weapon.

(6) The firer holds the weapon in the raised pistol position with the slide to the rear. He returns to the starting point and places the weapon on the stand. He turns in any excess ammunition to the ammunition point. On hearing the order to do so, he moves to the firing line.

f. **Table VI–Day Standing, CBRN**. All firers will wear protective masks with hoods. For this table, the firer receives one magazine containing seven rounds. Five targets (three single and one set of two) are exposed. Each is fired after the firer rotates to another firing point.

g. **Table VII–Night Standing**. For this table, the firer receives one magazine containing five rounds. Four targets (two single and one set of two) are exposed, starting with the same position used in the previous tables.

NOTE: Commanders may use the Engagement Skills Trainer (EST) 2000 to conduct Firing Tables VI and VII (CBRN and night fire).

A-3. Tower Operator's Authority

The tower operator is responsible for the range. For this reason, only he can give orders to scorers and firers on the range.

A-4. Conduct of Fire by Firing Table

For each table, the tower operator has scorers issue only the rounds required for that table. The following fire commands show how the tower operator runs range fire on the CPQC:

a. **Table I–Day Standing**. The tower operator orders firers to move to the firing line in preparation for firing. He orders the firers to position themselves next to the weapon stands and secure their weapons. On command, the scorer issues to the firer one magazine containing seven rounds.

(1) The tower operator commands–

TABLE ONE, STANDING POSITION, SEVEN ROUNDS.
LOAD AND LOCK.
READY ON THE RIGHT.
READY ON THE LEFT.
READY ON THE FIRING LINE.
UNLOCK YOUR WEAPONS.
FIRERS, WATCH YOUR LANE.

(2) The tower operator exposes the targets to the firers. When all targets have been exposed and engaged or lowered, the tower operator commands–

CEASE FIRE.
ARE THERE ANY ALIBIS?
(Alibis are given 10 seconds for each round not fired.)
CLEAR ALL WEAPONS.
CLEAR ON THE RIGHT.
CLEAR ON THE LEFT.
THE FIRING LINE IS CLEAR.
FIRERS, PLACE YOUR WEAPONS ON THE STANDS WITH SLIDES LOCKED TO THE REAR.
FIRERS AND SCORERS, MOVE DOWNRANGE AND CHECK YOUR TARGETS.
MARK AND COVER ALL HOLES.

b. Table II–Day Standing. The tower operator orders firers to secure their weapons. On command, the scorer issues to the firer one magazine containing a single round and another magazine containing seven rounds.

(1) The tower operator commands–

TABLE TWO, STANDING POSITION, EIGHT ROUNDS.
LOAD AND LOCK ONE MAGAZINE WITH ONE ROUND.
LOAD YOUR SEVEN-ROUND MAGAZINE WITHOUT COMMAND.
READY ON THE RIGHT.
READY ON THE LEFT.
READY ON THE FIRING LINE.
UNLOCK YOUR WEAPONS.
FIRERS, WATCH YOUR LANES.

(2) The tower operator exposes the targets to the firers. When all targets have been exposed and engaged or lowered, the tower operator commands–

CEASE FIRE.
ARE THERE ANY ALIBIS? (Alibis are given 10 seconds for each round not fired.)
CLEAR ALL WEAPONS.
CLEAR ON THE RIGHT.
CLEAR ON THE LEFT.
THE FIRING LINE IS CLEAR.

FIRERS, PLACE YOUR WEAPONS ON THE STAND WITH SLIDES LOCKED TO THE REAR.
FIRERS AND SCORERS, MOVE DOWNRANGE AND CHECK YOUR TARGETS.
MARK AND COVER ALL HOLES.

c. **Table III–Day Standing**. The tower operator orders the firers to secure their weapons. On command, the scorer issues to the firer one magazine containing seven rounds.

(1) The tower operator commands–

TABLE THREE, STANDING POSITION, SEVEN ROUNDS.
LOAD AND LOCK.
READY ON THE RIGHT.
READY ON THE LEFT.
READY ON THE FIRING LINE.
UNLOCK YOUR WEAPONS.
FIRERS, WATCH YOUR LANES.

(2) The tower operator exposes the targets to the firers. When all targets have been exposed and engaged or lowered, the tower operator commands–

CEASE FIRE.
ARE THERE ANY ALIBIS? (Alibis are given 10 seconds for each round not fired.)
CLEAR ALL WEAPONS.
CLEAR ON THE RIGHT.
CLEAR ON THE LEFT.
THE FIRING LINE IS CLEAR.
FIRERS, PLACE YOUR WEAPONS ON THE STAND WITH SLIDES LOCKED TO THE REAR.
FIRERS AND SCORERS, MOVE DOWNRANGE AND CHECK YOUR TARGETS.
MARK AND COVER ALL HOLES.

d. **Table IV–Day Standing**. The tower operator orders the firers to secure their weapons. On command, the scorer issues to the firer one magazine containing five rounds.

(1) The tower operator commands–

TABLE FOUR, STANDING POSITION, FIVE ROUNDS.
LOAD AND LOCK.
READY ON THE RIGHT.
READY ON THE LEFT.
READY ON THE FIRING LINE.
UNLOCK YOUR WEAPONS.
FIRERS, WATCH YOUR LANES.

(2) The tower operator exposes the targets to the firers. When all targets have been exposed and engaged or lowered, the tower operator commands–

CEASE FIRE.
ARE THERE ANY ALIBIS? (Alibis are given 10 seconds for each round not fired)

CLEAR ALL WEAPONS.
CLEAR ON THE RIGHT.
CLEAR ON THE LEFT.
THE FIRING LINE IS CLEAR.
FIRERS, PLACE YOUR WEAPONS ON THE STAND WITH SLIDES LOCKED TO THE REAR.
FIRERS AND SCORERS, MOVE DOWNRANGE AND CHECK YOUR TARGETS.
MARK AND COVER ALL HOLES.

e. **Table V–Day Moving Out**. The tower operator orders the firers to secure their weapons and move to the center of the trail 10 meters behind the firing line. On command, the scorer issues to the firer one magazine containing one round; a second magazine containing seven rounds; and a third magazine containing five rounds.

(1) The tower operator commands–

TABLE FIVE, STANDING POSITION, THIRTEEN ROUNDS.
LOAD AND LOCK ONE MAGAZINE WITH ONE ROUND.
READY ON THE RIGHT.
READY ON THE LEFT.
READY ON THE FIRING LINE.
PISTOLS AT THE READY POSITION.
UNLOCK YOUR WEAPON.
FIRERS, WATCH YOUR LANES.
MOVE OUT.

(2) The tower operator exposes the targets to the firers. After each target or group of targets has been engaged, he commands–

WEAPONS AT THE READY POSITION.
MOVE OUT.

(3) After the firers complete Table V, the tower operator commands–

CEASE FIRE.
ARE THERE ANY ALIBIS? (Alibis are given 10 seconds for each round not fired)
CLEAR ALL WEAPONS.
CLEAR ON THE RIGHT.
CLEAR ON THE LEFT.
THE FIRING LINE IS CLEAR.
FIRERS, KEEP YOUR WEAPONS UP AND DOWN RANGE.
SCORERS AND FIRERS MOVE BACK TO THE FIRING LINE AND PLACE YOUR WEAPONS ON THE STAND WITH SLIDES LOCKED TO THE REAR.
FIRERS AND SCORERS, MOVE DOWNRANGE AND CHECK YOUR TARGETS.
MARK AND COVER ALL HOLES.

(4) The tower operator has each scorer total the firer's scorecard and turn it in to the range officer or his

representative. The firing orders are rotated and the above sequence continued until all orders have fired.

f. **Table VI–Day Standing, CBRN**. The firer will wear a protective mask with hood.

(1) The tower operator orders the firers to position themselves next to the weapon stands. On command, the scorer issues to the firer one magazine containing seven rounds. The firer must get three hits to receive a "GO" on this table. The tower operator commands–

TABLE SIX, CBRN FIRE, STANDING POSITION, SEVEN ROUNDS.
LOAD AND LOCK.
READY ON THE RIGHT.
READY ON THE LEFT.
READY ON THE FIRING LINE.
UNLOCK YOUR WEAPONS.
FIRERS, WATCH YOUR LANES.

(2) The tower operator exposes the targets to the firers. When all targets have been exposed and engaged or lowered, the tower operator commands–

CEASE FIRE.
ARE THERE ANY ALIBIS? (Alibis are given 10 seconds for each round not fired)
CLEAR ALL WEAPONS.
CLEAR ON THE RIGHT.
CLEAR ON THE LEFT.
THE FIRING LINE IS CLEAR.
FIRERS, PLACE YOUR WEAPONS ON THE STAND. WITH SLIDES LOCKED TO THE REAR.
FIRERS AND SCORERS, MOVE DOWNRANGE AND CHECK YOUR TARGETS.
MARK AND COVER ALL HOLES.

g. **Table VII–Night Standing**. The tower operator orders the firers to position themselves next to the weapon stands. On command, the scorer issues to the firer one magazine containing five rounds. The firer must get two hits to receive a "GO" on this table.

(1) The tower operator commands–

TABLE SEVEN, NIGHT FIRE, STANDING POSITION, FIVE ROUNDS.
LOAD AND LOCK.
READY ON THE RIGHT.
READY ON THE LEFT.
READY ON THE FIRING LINE.
UNLOCK YOUR WEAPONS.
FIRERS, WATCH YOUR LANES.

(2) The tower operator exposes the targets to the firers. When all targets have been exposed and engaged or lowered, The tower operator commands–

CEASE FIRE.
ARE THERE ANY ALIBIS? (Alibis are given 10 seconds for each round not fired)
CLEAR ALL WEAPONS.
CLEAR ON THE RIGHT.
CLEAR ON THE LEFT.
THE FIRING LINE IS CLEAR.
FIRERS, PLACE YOUR WEAPONS ON THE STAND WITH SLIDES LOCKED TO THE REAR.
FIRERS AND SCORERS, MOVE DOWNRANGE AND CHECK YOUR TARGETS.
MARK AND COVER ALL HOLES.

A-5. Alibis

Alibis are fired after each table and where they occurred. Firers are allowed 10 seconds for each alibi. The same fire commands apply to alibis. If a weapon or target malfunctions while the firer is firing from a stationary position, he reports the malfunction. He keeps his weapon pointed up and downrange. Should the malfunction occur during Table V, the firer keeps his weapon pointed up and downrange, but he continues to move forward, keeping himself aligned with the firers to his right and left.

A-6. Rules

Certain rules apply to the conduct of fire during the CPQC:

a. **Assistance**. During instructional fire, the coach and assistant instructors should assist the firer in correcting errors. However, during record fire, no one may help or try to help the firer while or after he takes his position at the firing point.

b. **Accidental Discharges**. After the firer takes his place on the firing lane, every shot counts. Even if he fires away from the target or discharges the weapon accidentally, then that counts as his shot. He receives no replacement round or second chance.

c. **Fire on the Wrong Target**. Each firer observes the location of the target in his own lane. Shots fired on the wrong target count as a miss. A firer is credited only for the targets he hits in his own firing lane.

d. **Fire After the Signal to Lower Targets**. Any shot after the target starts to lower is scored as a miss.

e. **Extra Shot Fired at an E-Type Silhouette Target**. If the firer hits the target while the target is exposed, that is, before it begins to lower, then he receives credit for the hit. The number of rounds fired to obtain the hit does not matter.

f. **Excess Ammunition**. At the end of each firing table, the firer turns in any excess ammunition. This ammunition is not re-issued to him for use in the other firing tables.

g. **Target Sequence**. The tower operator sets a common target sequence for all lanes. This keeps a firer from getting ahead of the firers in adjacent lanes. Target sequence varies in distance from the firer. It starts with 31 meters and allows for no more than two 7-meter targets.

A-7. Scorecard

Figure A-1 shows an example completed DA Form 88-R (*Combat Pistol Qualification Course Scorecard*), and a blank copy is provided in the back of the book. The blank form may be reproduced locally on 8 1/2- by 11-inch paper. It may also be downloaded from the Internet at Army Knowledge Online (http://www.army.mil/usapa/eforms/). The scorecard lists the standards and provides scoring grids for the CPQC.

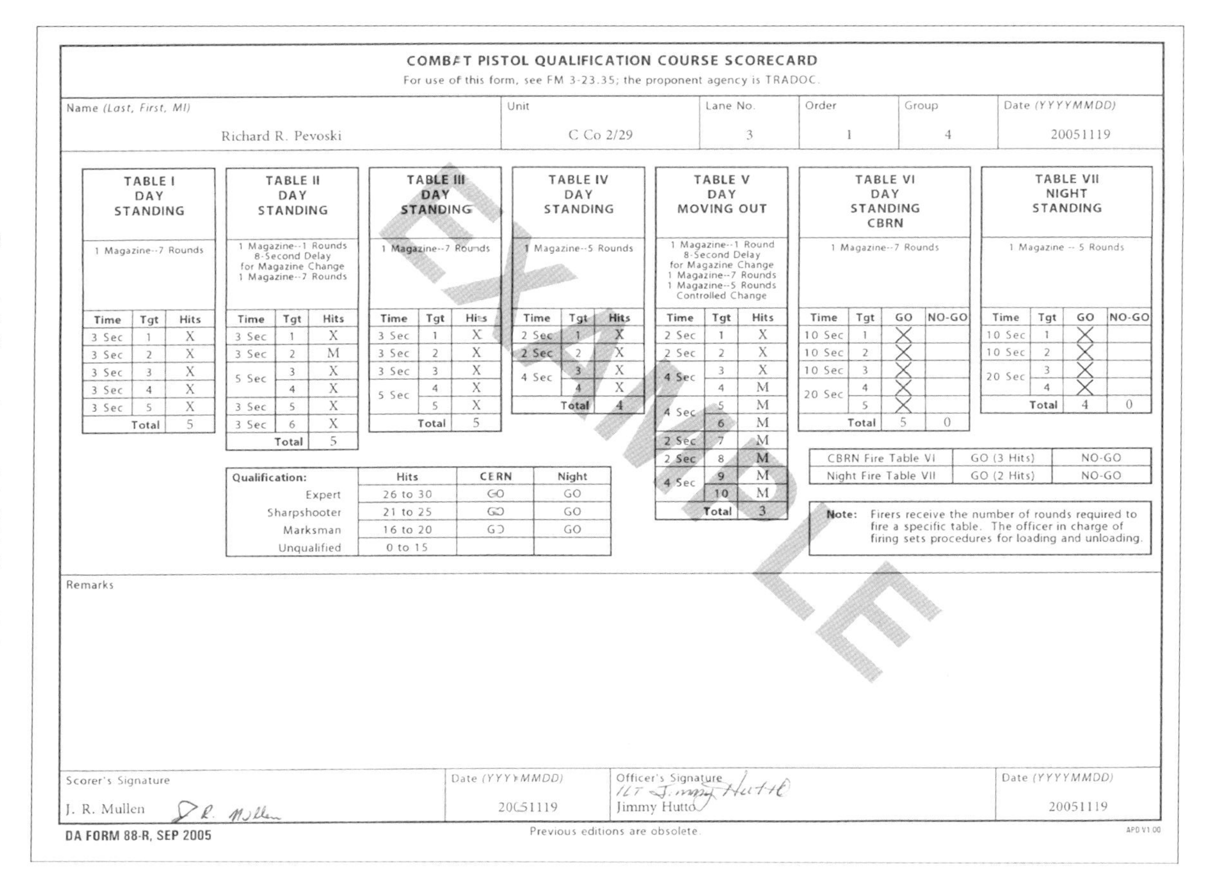

COMBAT PISTOL QUALIFICATION COURSE SCORECARD

For use of this form, see FM 3-23.35; the proponent agency is TRADOC.

Name (Last, First, MI)	Unit	Lane No.	Order	Group	Date (YYYYMMDD)
Richard R. Pevoski	C Co 2/29	3	1	4	20051119

TABLE I DAY STANDING — 1 Magazine--7 Rounds

Time	Tgt	Hits
3 Sec	1	X
3 Sec	2	X
3 Sec	3	X
3 Sec	4	X
3 Sec	5	X
	Total	5

TABLE II DAY STANDING — 1 Magazine--1 Rounds 8-Second Delay for Magazine Change 1 Magazine--7 Rounds

Time	Tgt	Hits
3 Sec	1	X
3 Sec	2	M
5 Sec	3	X
	4	X
3 Sec	5	X
3 Sec	6	X
	Total	5

TABLE III DAY STANDING — 1 Magazine--7 Rounds

Time	Tgt	Hits
3 Sec	1	X
3 Sec	2	X
3 Sec	3	X
5 Sec	4	X
	5	X
	Total	5

TABLE IV DAY STANDING — 1 Magazine--5 Rounds

Time	Tgt	Hits
2 Sec	1	X
2 Sec	2	X
4 Sec	3	X
	4	X
	Total	4

TABLE V DAY MOVING OUT — 1 Magazine--1 Round 8-Second Delay for Magazine Change 1 Magazine--7 Rounds 1 Magazine--5 Rounds Controlled Change

Time	Tgt	Hits
2 Sec	1	X
2 Sec	2	X
4 Sec	3	X
	4	M
4 Sec	5	M
	6	M
2 Sec	7	M
2 Sec	8	M
4 Sec	9	M
	10	M
	Total	3

TABLE VI DAY STANDING CBRN — 1 Magazine--7 Rounds

Time	Tgt	GO	NO-GO
10 Sec	1	X	
10 Sec	2	X	
10 Sec	3	X	
20 Sec	4	X	
	5	X	
	Total	5	0

TABLE VII NIGHT STANDING — 1 Magazine -- 5 Rounds

Time	Tgt	GO	NO-GO
10 Sec	1	X	
10 Sec	2	X	
20 Sec	3	X	
	4	X	
	Total	4	0

CBRN Fire Table VI	GO (3 Hits)	NO-GO
Night Fire Table VII	GO (2 Hits)	NO-GO

Note: Firers receive the number of rounds required to fire a specific table. The officer in charge of firing sets procedures for loading and unloading.

Qualification:	Hits	CERN	Night
Expert	26 to 30	GO	GO
Sharpshooter	21 to 25	GO	GO
Marksman	16 to 20	GO	GO
Unqualified	0 to 15		

Remarks

Scorer's Signature	Date (YYYYMMDD)	Officer's Signature	Date (YYYYMMDD)
J. R. Mullen	20051119	1LT Jimmy Hutto Jimmy Hutto	20051119

DA FORM 88-R, SEP 2005 — Previous editions are obsolete. — APD V1.00

Figure A-1. Example completed DA Form 88-R.

NOTE: Numbers in the "Tgt" column do not represent a particular sequence in which the targets will appear. They just identify how many targets the firer will engage in each firing table.

a. Each time a firer hits or kills a target, the scorer places an "X" in the "Hits" column. Each hit is worth one point. After the firer finishes firing, the scorer totals and signs the scorecard.

b. The following qualification standards are also shown on the form:
   - Expert–26 to 30 hits.
   - Sharpshooter–21 to 25 hits.
   - Marksman–16 to 20 hits.
   - Unqualified–0 to 15 hits.

c. The CBRN and night firing tables are scored as GO or NO-GO. The firer either qualifies on those tables, or not. For each of these tables (VI and VII), the firer gets a GO if he hits the target, and a NO-GO if he misses.

d. To qualify, the firer must earn a minimum total score of 16 on Tables I through V and must receive three hits on Table VI (CBRN) and two hits on Table VII (Night).

A-8. Targets

Each firing lane requires seven electrical, device-type targets as well as a single E-type silhouette. Aggressor figures may be superimposed on the silhouettes to add realism to the course of fire.

A-9. Quick-Fire Target Training Device

The unit may procure a quick-fire target training device (QTTD) locally. To ensure standardization, quality, durability, and appearance, the device should be constructed by a qualified organization with documented experience producing similar devices such as the training aids section of the local Training Support Center.

APPENDIX B*

ALTERNATE PISTOL QUALIFICATION COURSE

Once the Soldier completes instructional fire, he must complete the Combat Pistol Qualification Course (CPQC) for the record. However, when the CPQC is unavailable, the Alternate Pistol Qualification Course (APQC) may be used.
The tower operator is completely responsible for and in charge of the range and the course. He controls absolutely all activities related to firing. The tower operator tells the scorers what to do when, for example, when to issue the preloaded magazines to firers. Only the tower operator may issue the order to fire. Scorers and firers must await the tower operator's orders.

B-1. Conditions and Standards

The firer is given 40 rounds of ammunition for Tables I through IV and 14 rounds for Tables V and VI:

a. **Table I–Day Standing** Within 21 seconds, engage the 25-meter APQC target from the standing position with 7 rounds of ammunition; given one 7-round magazine during daylight hours.
b. **Table II–Day Kneeling**. Within 45 seconds, engage the 25-meter APQC target from the kneeling position with 13 rounds during hours of daylight. From a standing position, assume a good kneeling position, engage the target with 6 rounds, perform a rapid magazine change, and engage the target with 7 rounds.
c. **Table III–Day Crouching**. Within 35 seconds, engage the 25-meter APQC target from the crouching position with 10 rounds; given two magazines with 5 rounds each during daylight hours. From a standing position, assume a good crouching position, engage the target with one 5-round magazine, perform a rapid magazine change, and engage the target with the second 5-round magazine.

d. **Table IV–Day Prone**. Within 35 seconds, engage the 25-meter APQC target from the prone position with 10 rounds; given two magazines with 5 rounds each during daylight hours. From a standing position, assume a good prone position, engage the target with one 5-round magazine, perform a rapid magazine change, and engage the target with the second 5-round magazine.

e. **Table V–Day CBRN Crouching.** Within 70 seconds, engage the 25-meter target from the crouching position with 7 rounds; given one 7-round magazines under simulated CBRN conditions.

f. **Table VI–Night Crouching.** Within 70 seconds, engage a 25-meter target from a crouching position with 7 rounds; given one 7-rounds magazine under night conditions.

B-2. Conduct of Fire

Commands shape the conduct of range fire in the APQC. When the firer uses a 9-mm pistol, he fires the first round in each table in double-action mode. At the end of each firing table, he returns excess ammunition to the scorer. He may not use it in other firing tables. At the end of the course, the scorer returns all excess ammunition to the ammunition point.

NOTES: 1. Commanders may use the Engagement Skills Trainer (EST) 2000 for Tables V (Day CBRN Crouching) and VI (Night Crouching).
2. Only the tower operator may give firing instructions.

a. **Table I–Day Standing**. The tower operator gives the order to move to the firing line and to prepare to fire. On the tower operator's command, the scorer issues to the firer one magazine containing seven rounds.

(1) The tower operator then commands–

TABLE ONE, STANDING POSITION, SEVEN ROUNDS.
LOAD AND LOCK. ONE SEVEN ROUND MAGAZINE.
IS THE FIRING LINE READY? (Firers using 9-mm pistols place them on double-action.)
READY ON THE RIGHT.
READY ON THE LEFT.
THE FIRING LINE IS READY.
FIRERS, UNLOCK YOUR WEAPONS.
FIRERS, WATCH YOUR LANE!

(2) At the end of the prescribed firing time, the tower operator commands–

CEASE FIRE.
ARE THERE ANY ALIBIS? (Alibis get eight seconds for each round not fired.)
UNLOAD AND CLEAR ALL WEAPONS.
IS THE FIRING LINE CLEAR?
CLEAR ON THE RIGHT.
CLEAR ON THE LEFT.
THE FIRING LINE IS NOW CLEAR.
FIRERS, PLACE YOUR WEAPON ON THE STANDS WITH SLIDES LOCKED TO THE REAR.
FIRERS AND SCORERS, MOVE DOWNRANGE AND CHECK YOUR TARGETS.
MARK AND COVER ALL HOLES.

b. **Table II–Day Kneeling**. The tower operator orders firers to move up to the firing line. On the tower operator's command, the scorer issues two magazines, one loaded with six rounds and the other with seven, to the firer.

(1) The tower operator then commands–

TABLE TWO, KNEELING POSITION, SIX ROUNDS.
LOAD AND LOCK ONE SIX-ROUND MAGAZINE.
LOAD YOUR SEVEN-ROUND MAGAZINE WITHOUT COMMAND.
IS THE FIRING LINE READY? (Firers using 9-mm pistols place them on double action.)
READY ON THE LEFT.
READY ON THE RIGHT.
THE FIRING LINE IS READY.
FIRERS, UNLOCK YOUR WEAPONS.
FIRERS, WATCH YOUR LANES.

(2) The tower operator then commands–

CEASE FIRE.
ARE THERE ANY ALIBIS? (Alibis get eight seconds for each round not fired.)
UNLOAD AND CLEAR ALL WEAPONS.
IS THE FIRING LINE CLEAR?
CLEAR ON THE RIGHT
CLEAR ON THE LEFT
THE FIRING LINE IS NOW CLEAR.
FIRERS, PLACE YOUR WEAPON ON THE STANDS WITH SLIDES LOCKED TO THE REAR.
FIRERS AND SCORERS MOVE DOWNRANGE AND CHECK YOUR TARGETS.
MARK AND COVER ALL HOLES.

c. **Table III–Day Crouching**. The tower operator orders firers to move to the firing line.

(1) On the tower operator's command, the scorer issues two five-round magazines to the firer:

TABLE THREE, CROUCHING POSITION, FIVE ROUNDS.
LOAD AND LOCK ONE FIVE-ROUND MAGAZINE.
IS THE FIRING LINE READY? (Firers using 9-mm pistols place them on double action.)
READY ON THE RIGHT.
READY ON THE LEFT.
THE FIRING LINE IS READY.
FIRERS, WATCH YOUR LANES.

(2) At the end of the prescribed firing time, the tower operator commands–

CEASE FIRE.
ARE THERE ANY ALIBIS? (Alibis are given 10 seconds for each round not fired.)
UNLOAD AND CLEAR ALL WEAPONS.
IS THE FIRING LINE CLEAR?
CLEAR ON THE RIGHT.
CLEAR ON THE LEFT.
THE FIRING LINE IS NOW CLEAR.
FIRERS, PLACE YOUR WEAPON ON THE STAND WITH SLIDES LOCKED TO THE REAR.
FIRERS AND SCORERS, MOVE DOWNRANGE AND CHECK YOUR TARGETS.
MARK AND COVER ALL HOLES.

d. **Table IV–Day Prone Unsupported**. The tower operator orders firers to move to the firing line. On the tower operator's command, the scorer issues two five-round magazines to the firer.

(1) After the firer completes Table IV, the scorer and firer repair or replace targets for the next firing order:

TABLE FOUR, PRONE POSITION, FIVE ROUNDS.
LOAD AND LOCK ONE FIVE-ROUND MAGAZINE.
LOAD YOUR SECOND FIVE-ROUND MAGAZINE WITHOUT COMMAND.
IS THE FIRING LINE READY?
READY ON THE RIGHT.
READY ON THE LEFT.
THE FIRING LINE IS READY.
FIRERS, UNLOCK YOUR WEAPONS.
FIRERS, WATCH YOUR LANES.

(2) At the end of the prescribed firing time, the tower operator commands–

CEASE FIRE.
ARE THERE ANY ALIBIS? (Alibis are given 10 seconds for each round not fired.)
UNLOAD AND CLEAR ALL WEAPONS.
IS THE FIRING LINE CLEAR?

CLEAR ON THE RIGHT.
CLEAR ON THE LEFT.
THE FIRING LINE IS NOW CLEAR.
FIRERS, PLACE YOUR WEAPON ON THE STAND WITH SLIDES LOCKED TO THE REAR.
FIRERS AND SCORERS, MOVE DOWN RANGE AND CHECK YOUR TARGETS.
MARK AND COVER ALL HOLES.

e. **Table V–Day CBRN Crouching.** All firers wear protective masks with hoods. The tower operator orders firers to move to the firing line. On the tower operator's command, the scorer issues one seven-round magazine to the firer.

(1) Again, he issues the same commands he did for Table I.

TABLE FIVE, CBRN FIRE, CROUCHING POSITION, SEVEN ROUNDS.
LOAD AND LOCK ONE MAGAZINE.
IS THE FIRING LINE READY? (Firers using 9-mm pistols place them on double-action.)
READY ON THE RIGHT.
READY ON THE LEFT.
THE FIRING LINE IS READY.
FIRERS, WATCH YOUR LANES.

(2) At the end of the prescribed firing time, the tower operator commands–

CEASE FIRE.
ARE THERE ANY ALIBIS? (Alibis are given 10 seconds for each round not fired.)
UNLOAD AND CLEAR ALL WEAPONS.
IS THE FIRING LINE CLEAR?
READY ON THE RIGHT.
READY ON THE LEFT.
THE FIRING LINE IS NOW CLEAR.
FIRERS, PLACE YOUR WEAPON ON THE STAND WITH SLIDES LOCK TO THE REAR FIRERS AND SCORERS, MOVE DOWNRANGE AND CHECK YOUR TARGETS.
MARK AND COVER ALL HOLES.

f. **Table VI–Night Crouching.** The tower operator orders firers to move to the firing line and to prepare to fire. On the tower operator's command, the scorer issues one seven-round magazine to the firer.

(1) The tower operator commands–

TABLE SIX, NIGHT FIRE, CROUCHING POSITION, SEVEN ROUNDS.
LOAD AND LOCK ONE MAGAZINE.
IS THE FIRING LINE READY? (Firers using 9-mm pistols place them on double-action.)
READY ON THE RIGHT.
READY ON THE LEFT.
THE FIRING LINE IS READY.

FIRERS, UNLOCK YOUR WEAPONS.
FIRERS, WATCH YOUR LANES.

(2) At the end of the prescribed firing time, the tower operator commands–

CEASE FIRE.
ARE THERE ANY ALIBIS? (Alibis get 8 seconds for each round not fired.)
UNLOAD AND CLEAR ALL WEAPONS.
IS THE FIRING LINE CLEAR?
READY ON THE RIGHT.
READY ON THE LEFT.
THE FIRING LINE IS NOW CLEAR.
FIRERS, PLACE YOUR WEAPONS ON THE STANDS WITH SLIDES LOCKED TO THE REAR.
FIRERS AND SCORERS, MOVE DOWNRANGE AND CHECK YOUR TARGETS.
MARK AND COVER ALL HOLES.

B-3. Alibis

The scorer reports and records any weapon or target malfunction that occurs during fire. The firer is allowed one alibi at the end of each table. For Tables I through IV, he is allowed 8 seconds for each alibi. For Tables V and VI, he is allowed 10 seconds each. All alibis are fired where the malfunction occurred, and the same firing commands are used.

B-4. Scorecard

Figure B-1 shows a 25–meter E-type silhouette. Figure B-2 shows an example completed DA Form 5704-R (*Alternate Pistol Qualification Course Scorecard*). A blank copy is provided in the back of the book. The blank form may be reproduced locally on 8 1/2- by 11-inch paper. It may also be downloaded from the Internet at Army Knowledge Online (http://www.army.mil/usapa/eforms/).

a. **Firing Tables I thru IV**. Each time a firer hits or kills a target, the scorer places an "X" (for a hit) or "M" (for a miss) in the appropriate box, then writes the number of each in the "Hits" and "Misses" columns to the right. After the firer completes the first four firing tables, the scorer tallies the total hits and misses and uses the scoring grid to determine if the firer qualified on those four tables.

EXPERT–36 to 40 hits.

SHARPSHOOTER–30 to 35 hits.
MARKSMAN–24 to 29 hits.
UNQUALIFIED–0 to 23 hits.

b. **Firing Tables V and VI**. Each time a firer hits or kills a target, the scorer places an "X" (hit) or "M" (miss) in the appropriate box. Then, he writes the total number of hits and misses in the columns to the right. The firer must hit four targets in each table in order to receive a "GO" on that table.

B-5. Assistance

During instructional fire, the coach and assistant instructors should help the firer correct errors. However, during record fire, no one may help or try to help the firer while or after he takes his position at the firing point.

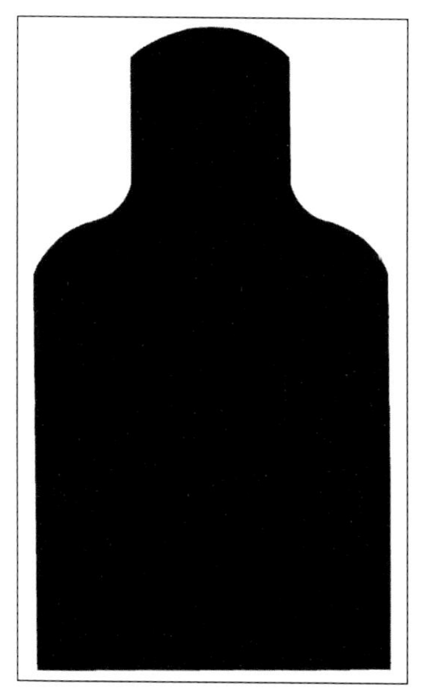

Figure B-1. E-type silhouette, 25-meter, without rings.

ALTERNATE PISTOL QUALIFICATION COURSE SCORECARD
For use of this form, see FM 3-23.35; the proponent agency is TRADOC.

Name *(First, Last, MI)*	Unit	Lane No.	Order	Date *(YYYYMMDD)*
Richard P. Pevoski	C Co, 2/29	1	3	20051119

TABLES I THRU IV -- DAY Scorer marks each hit with a "X" and each miss with an "M". He writes the totals for each in the "Hits" and "Misses" columns. When the firer completes Tables I thru IV, the scorer totals the columns and uses the scoring grid below. The firer must achieve at least 24 hits to qualify.	**HITS**	**MISSES**
TABLE I: DAY STANDING Magazines: 1 Rounds: 7 Time: 21 seconds **Hits:** X X X X X X X	7	0
TABLE II: DAY KNEELING Magazines: 2 Rounds: 6 in one magazine, 7 in the other Time: 45 seconds **Hits:** X X X X X X X X X X X X M	12	1
TABLE III: DAY CROUCHING Magazines: 2 Rounds: 5 in each magazine Time: 35 seconds **Hits:** X X X X X X X X X X	10	0
TABLE IV: DAY PRONE UNSUPPORTED Magazines: 2 Rounds: 5 in each magazine Time: 35 seconds **Hits:** X X X X X X X X X X	10	0
SCORING GRID FOR TABLES I THRU IV Expert 36 to 40; Sharpshooter 30 to 35; Marksman 24 to 29; Not Qualified 23 or less — **Hit/Miss Totals:**	39	1
TABLES V AND VI - LIMITED VISIBILITY After firer shoots each table, the scorer marks the appropriate block.	**GO**	**NO-GO**
TABLE V: DAY CBRN CROUCHING Magazines: 1 Rounds: 7 Time: 70 seconds **Hits:** X X X X X X X (Four hits are required for a GO.)	7	0
TABLE VI: NIGHT CROUCHING Magazines: 1 Rounds: 7 Time: 70 seconds **Hits:** X X X X X X X (Four hits are required for a GO.)	7	0

Remarks

Scorer's Signature	Date *(YYYYMMDD)*	Officer's Signature	Date *(YYYYMMDD)*
SSG Katisha Ann Duncan	20051119	CPT Rolando O'Reilly	20051119

DA FORM 5704-R, SEP 2005 Previous editions are obsolete. APD V1.00

Figure B-2. Example completed DA Form 5704-R.

APPENDIX C

TRAINING SCHEDULES

To aid in the individual training phase, training schedules for the courses in pistol marksmanship training are described in this appendix. These schedules are based on the desirable number of training hours for a pistol course. They should be used as a guide in preparing lesson plans; conditions may require a longer or shorter period to complete the training. When time is available, additional training should be included in the schedule. When suggested equipment and training aids are not available, the best that are available should be improvised or substituted. Each firer should be allowed 50 rounds for instructional firing and 40 rounds for record firing.

C-1. 9-MM SEMIAUTOMATIC PISTOL, PRACTICE OR INSTRUCTIONAL FIRING COURSE (12 HOURS)

Period	Hours		Lesson	References	Training Facilities	Training Aids
	Peace	Mobilization				
MECHANICAL TRAINING (4 Hours)						
1	2	2	Characteristics, disassembly and assembly, functioning, and care and cleaning.	TM 9-1005-317-10, TM 9-1005 325-10.	Classroom or field.	For each instructor: chalkboard, working model, projector and screen. For each man: cleaning equipment. For each group: table or suitable ground cloth.
2	2	2	Malfunctions, stoppages, immediate action, loading, unloading ammunition, and safety precautions.	TM 9-1005-317-10, TM 9-1005-325-10.	....do....	Same as period 1 plus ammunition display.

C-1. 9-MM SEMIAUTOMATIC PISTOL, PRACTICE OR INSTRUCTIONAL FIRING COURSE (12 HOURS)—*(Continued)*

Period	Hours		Lesson	References	Training Facilities	Training Aids
	Peace	Mobilization				
			PREPARATORY MARKSMANSHIP TRAINING (6 Hours)			
3	3	3	Coaching, aiming, grip, positions, trigger squeeze (to include double-action), target engagement, and slow-fire exercise.	Chapter 2 of this manual.	....do....	For each man: one pistol with magazine. For all: E-silhouette.
			RANGE FIRING (2 Hours)			
4	2	2	Instructional Firing Tables I through V, Combat Pistol Qualification Course.	Appendix A of this manual.	Live-fire range.	Equipment used in period 6 of the qualification course.

C-2. 9-MM SEMIAUTOMATIC PISTOL, QUALIFICATION COURSE (12 HOURS)

Period	Hours		Lesson	References	Training Facilities	Training Aids
	Peace	Mobilization				
			MECHANICAL TRAINING (4 Hours)			
1	2	2	Characteristics, disassembly and assembly, functioning, and care and cleaning.	TM 9-1005-317-10, TM 9-1005-325-10.	Classroom or field.	For each instructor: chalkboard, working model, projector and screen. For each man: cleaning equipment. For each group: table or suitable ground cloth.

2	2	2	Malfunctions, stoppages, immediate action, loading, unloading, ammunition, and safety precautions.	TM 9-1005-317-10, TM 9-1005-325-10.	....do....	Same as period 1 plus ammunition display.

C-2. 9-MM SEMIAUTOMATIC PISTOL, QUALIFICATION COURSE (12 HOURS) *—(Continued)*

Period	Hours		Lesson	References	Training Facilities	Training Aids
	Peace	Mobilization				
PREPARATORY MARKSMANSHIP TRAINING (4 Hours)						
3	2	2	Coaching, aiming, grip, positions, trigger squeeze (to include double-action), target engagement, and slow-fire exercise.	Chapter 2 of this manual.	....do....	For each man: one pistol with magazine, For all: E-silhouette.
4	2	2	Review and examination.	All previous references.	....do....	For all: all equipment used in previous periods.
RANGE FIRING (4 Hours)						
5	2	2	Instructional firing combat pistol qualification course, for practice with a coach or instructor.	Existing range regulations. Appendix A of this manual.	Pistol range.	For all: all equipment used for periods 3 and 4 plus scorecard and ammunition.
6	2	2	Record firing, Tables I through V, combat pistol qualification course.	Appendix A of this manual.	....do....	.....do....

Glossary

APQC	alternate pistol qualification course
AR	Army regulation
CBRN	chemical, biological, radiological, or nuclear
CLP	cleaner, lubricant, preservative
CPQC	combat pistol qualification course
CTA	common table of allowances
DA	Department of the Army
EENT	end evening nautical twilight
EMNT	end morning nautical twilight
FM	field manual
HQ	headquarters
LSA	lubricating [oil], semifluid, automatic [weapons]
MM	millimeter
MOPP	mission-oriented protective posture
NATO	North Atlantic Treaty Organization
NCOIC	noncommissioned officer in charge
NG	[Army] National Guard
OIC	officer in charge
QTTD	quickfire target-training device
RBC	rifle bore cleaner
RH	right hand
TM	technical manual
TRADOC	[US Army] Training and Doctrine Command
USAR	US Army Reserve

REFERENCES

DOCUMENTS NEEDED

These documents must be available to the intended users of this publication.

AR 385-63	Policies and Procedures for Firing Ammunition for Training, Target Practice, and Combat. 15 October 1983.
DA Form 88	Combat Pistol Qualification Course Scorecard.
DA Form 5704-R	Alternate Pistol Qualification Course.
TM 9-1005-317-10	Operator's Manual for Pistol, Semiautomatic, 9-mm, M9 (1005-01-118-2640). 31 July 1985.
TM 9-1005-325-10	Operator's Manual for Pistol, Compact, 9-mm, M11 (1005-01-336-8265) and Pistol, Compact, 9-mm, M11 with Tritium Sights (1005-01-340-0096). 16 December 1993.
TM 9-1300-200	Ammunition, General. 3 October 1969.

SOURCES USED

These are the sources quoted or paraphrased in this publication.

AR 140-1	Mission, Organization, and Training. 1 September 1994.
CTA 8-100	Army Medical Department Expendable/ Durable Items. 31 August 1994.
CTA 50-970	Expendable/Durable Items (Except: Medical, Class V, Repair Parts and Heraldic Items). 21 September 1990.
DA Pam 350-38	Standards in Weapons Training. 3 July 1997.
DA Pam 738-750	Functional Users' Manual for the Army Maintenance Management System (TAMMS). 1 August 1994.
FM 3-4	NBC Protection. 29 May 1992, with Changes 1-2, 28 October 1992-21 February 1996.
FM 3-5	NBC Decontamination. 28 July 2000.

FM 19-10	The Military Police Law and Order Operations. 30 September 1987.
FM 21-11	First Aid for Soldiers. 27 October 1988, with Changes 1-2, 28 August 1989-4 December 1991.
FM 21-75	Combat Skills of the Soldier. 3 August 1984.
TC 25-8	Training Ranges. 25 February 1992.
TM 9-1005-317-23&P	Unit and Intermediate Direct Support Maintenance Manual (Including Repair Parts and Special Tools List) for Pistol, Semiautomatic, 9-mm, M9 (1005-01-118-2640). 16 October 1987.
TM 9-1005-325-23&P	Unit and Intermediate Direct Support Maintenance Manual (Including Repair Parts and Special Tools List) for Pistol, Compact, 9-mm, M11 (1005-01-336-8265) and Pistol, Compact, 9-mm M11 with Tritium Sights (1005-01-340-0096). 1 March 1993.
TM 9-6920-210-14&P	Operator and General Support Maintenance Manual (Including Basic Issue Items List and Repair Parts List) for Small Arms Targets and Target Material. 2 June 1992.
TM 43-0001-27	Army Ammunition Data Sheets for Small Caliber Ammunition. 29 April 1994.

INTERNET SOURCES USED

The following Internet sources were used in the preparation of this manual.

U. S. Army Publishing Agency, http://www.usapa.army.mil

Army Doctrine and Training Digital Library, http://www.adtdl.army.mil

INDEX

air-operated pistol,
.177–mm, 42–44
alibis, 57–61, 67–70
alternate pistol qualification
course, 65–71
alibis, 70
conduct of fire, 66
form, 72
scoring, 70
tables, 70–71

basic marksmanship, 15–28
aiming, 19–20
breath control, 20–21
fundamentals, 15, 22, 24
positions, 23–29
target engagement 23
trigger squeeze, 21–22
ball-and-dummy method, 41
breath control, 20–21

calling the shot, 41
coaching, 40–41
combat marksmanship 29–40
nuclear, biological, chemical
firing, 40
poor visibility firing, 39–40
reloading techniques, 37–39
target engagement, 31
techniques of firing 29–31
traversing, 31–37
combat pistol qualification
course, 53–63
alibis, 61
conduct of fire, 56–61
rules, 61–62
scorecard, 62–63
tables, 54–56
targets, 64
combat reloading
techniques, 37–39
one-hand, 39
rapid, 38
tactical, 38

DA Form 88-R, Combat
Pistol Qualification Course
Scorecard, 62–63
DA Form 5704-R, Alternate Pistol
Qualification Course, 70, 72

equipment data, 2
pistol automatic, 9-mm,
M9, 2
pistol automatic, 9-mm,
M11, 2

flash sight picture, 29, 30
forms, reproducible, 83, 85

grip, 15–19
isometric tension, 18
one-hand, 15–16
two-hand, 16–19
fist, 16–17
palm-supported, 17
weaver, 18

hand-and-eye coordination, 29

instructional practice and record
qualification, 40–41

marksmanship training, 15–51
basic, 15–23
coaching and training
aids, 40–48
combat, 15–51

phases, 43, 50
safety, 48–51
malfunctions, 12–13

nuclear, biological, chemical firing, 40

pistols
M9 automatic 9-mm, 1–13
equipment data, 2
operation, 9–10
M11 automatic 9-mm, 1–13
equipment data, 2
operation, 9–10
point of aim, 18–19, 23, 24, 30
poor visibility firing, 2–13
positions, 22–28
crouch, 25–26
kneeling, 25
kneeling supported, 28–29
pistol–ready, 24
prone, 27
standing without support, 25
standing with support, 27–28

qualification courses
alternate pistol qualification course, 65–72
combat pistol qualification course, 53–64
qualification firing, 50–51
quick-fire point shooting 30
quick-fire sighting, 31
quick-fire target training device, 43, 64

range firing courses, 47–48
alternate pistol qualification course, 65–71
combat pistol qualification course, 53–64
reloading techniques, 37–39

safety, 48–51
scorecard, DA Form 88–R, Combat Pistol Qualification Course, 63
scorecard, DA Form 5704–R, Alternate Pistol Qualification Course, 72
slow-fire exercise, 42

target engagement, 23
recoil, 23
trigger jerk, 23
heeling, 23
targets, 54
techniques of firing, 29–31
training aids, 40–47
air–operated pistol, .177 mm, 42–43
quick–fire target training device, 43–46, 64
training schedules, 73–75
traversing, 31–37
trigger squeeze, 21–22

COMBAT PISTOL QUALIFICATION COURSE SCORECARD

For use of this form, see FM 3-23.35; the proponent agency is TRADOC.

Name *(Last, First, MI)*	Uni	Lane No.	Order	Group	Date *(YYYYMMDD)*

TABLE I DAY STANDING

1 Magazine--7 Rounds

Time	Tgt	Hits
3 Sec	1	
3 Sec	2	
3 Sec	3	
3 Sec	4	
3 Sec	5	
	Total	

TABLE II DAY STANDING

1 Magazine--1 Rounds 8-Second Delay for Magazine Change 1 Magazine--7 Rounds

Time	Tgt	Hits
3 Sec	1	
3 Sec	2	
5 Sec	3	
	4	
3 Sec	5	
3 Sec	6	
	Total	

TABLE III DAY STANDING

1 Magazine--7 Rounds

Time	Tgt	Hits
3 Sec	1	
3 Sec	2	
3 Sec	3	
5 Sec	4	
	5	
	Total	

TABLE IV DAY STANDING

1 Magazine--5 Rounds

Time	Tgt	Hits
2 Sec	1	
2 Sec	2	
4 Sec	3	
	4	
	Total	

TABLE V DAY MOVING OUT

1 Magazine--1 Round 8-Second Delay for Magazine Change 1 Magazine--7 Rounds 1 Magazine--5 Rounds Controlled Change

Time	Tgt	Hits
2 Sec	1	
2 Sec	2	
4 Sec	3	
	4	
4 Sec	5	
	6	
2 Sec	7	
2 Sec	8	
4 Sec	9	
	10	
	Total	

TABLE VI DAY STANDING CBRN

1 Magazine--7 Rounds

Time	Tgt	GO	NO-GO
10 Sec	1		
10 Sec	2		
10 Sec	3		
20 Sec	4		
	5		
	Total		

TABLE VII NIGHT STANDING

1 Magazine -- 5 Rounds

Time	Tgt	GO	NO-GO
10 Sec	1		
10 Sec	2		
20 Sec	3		
	4		
	Total		

Qualification:	Hits	CBRN	Night
Expert	26 to 30	GO	GO
Sharpshooter	21 to 25	GO	GO
Marksman	16 to 20	GO	GO
Unqualified	0 to 15		

CBRN Fire Table VI	GO (3 Hits)	NO-GO
Night Fire Table VII	GO (2 Hits)	NO-GO

Note: Firers receive the number of rounds required to fire a specific table. The officer in charge of firing sets procedures for loading and unloading.

Remarks

Scorer's Signature	Date *(YYYYMMDD*	Officer's Signature	Date *(YYYYMMDD)*

DA FORM 88-R, SEP 2005

Previous editions are obsolete.

APD V1.00

ALTERNATE PISTOL QUALIFICATION COURSE SCORECARD

For use of this form, see FM 3-23.35; the proponent agency is TRADOC.

Name *(First, Last, MI)*	Unit	Lane No.	Order	Date *(YYYYMMDD)*

TABLES I THRU IV -- DAY Scorer marks each hit with a "X" and each miss with an "M". He writes the totals for each in the "Hits" and "Misses" columns. When the firer completes Tables I thru IV, the scorer totals the columns and uses the scoring grid below. The firer must achieve at least 24 hits to qualify.	HITS	MISSES
TABLE I: DAY STANDING Magazines: 1 Rounds: 7 Time: 21 seconds **Hits:** ☐☐☐☐☐☐☐		
TABLE II: DAY KNEELING Magazines: 2 Rounds: 6 in one magazine, 7 in the other Time: 45 seconds **Hits:** ☐☐☐☐☐☐☐☐☐☐☐☐☐		
TABLE III: DAY CROUCHING Magazines: 2 Rounds: 5 in each magazine Time: 35 seconds **Hits:** ☐☐☐☐☐☐☐☐☐☐		
TABLE IV: DAY PRONE UNSUPPORTED Magazines: 2 Rounds: 5 in each magazine Time: 35 seconds **Hits:** ☐☐☐☐☐☐☐☐☐☐		
SCORING GRID FOR TABLES I THRU IV Expert: 36 to 40 Sharpshooter: 30 to 35 Marksman: 24 to 29 Not Qualified: 23 or less **Hit/Miss Totals:**		

TABLES V AND VI - LIMITED VISIBILITY After firer shoots each table, the scorer marks the appropriate block.	GO	NO-GO
TABLE V: DAY CBRN CROUCHING Magazines: 1 Rounds: 7 Time: 70 seconds **Hits:** ☐☐☐☐☐☐☐ (Four hits are required for a GO.)		
TABLE VI: NIGHT CROUCHING Magazines: 1 Rounds: 7 Time: 70 seconds **Hits:** ☐☐☐☐☐☐☐ (Four hits are required for a GO.)		

Remarks

Scorer's Signature	Date *(YYYYMMDD)*	Officer's Signature	Date *(YYYYMMDD)*

DA FORM 5704-R, SEP 2005 Previous editions are obsolete. APD V1.00

ALSO AVAILABLE

The Modern Day Gunslinger

The Ultimate Handgun Training Manual

by Don Mann

Foreword by Lt. Col. David Grossman

A result of twelve years of research, *The Modern Day Gunslinger* was written to meet the needs of the gun owner, the experienced shooter, those who own a weapon strictly for home and self-defense, and for the military member who wants to become a better shooter in defense of our country. It's also for the law enforcement officer who risks his or her life going against the thugs of our society and for anyone interested in learning the defensive and tactical training techniques from some of the best and most experienced shooters in the world.

The shooting skills taught in this book carry broad application in civilian, law enforcement, and military contexts. Common criminals, terrorists, assailants—the enemy and threat—all will find themselves outgunned in the face of a properly armed and trained gunslinger. Members of the armed services, government, and law enforcement agencies, as well as civilians, will find that the close-range shooting methods addressed in this book can provide a decisive advantage. An all-encompassing manual that addresses safety, equipment, tactics, and the best practices for all shooters, *The Modern Day Gunslinger* is the most complete book on shooting ever published. It's a book that, in the words of senior special agent and US government senior weapons and tactics instructor Dick Conger, "will save lives."

$17.95 Paperback • ISBN 978-1-60239-986-0

ALSO AVAILABLE

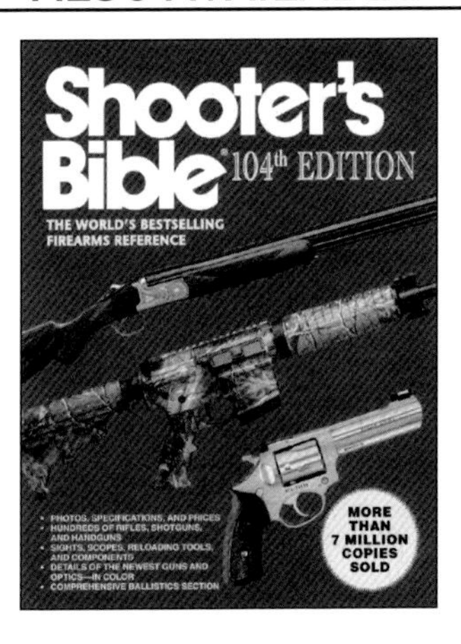

Shooter's Bible, 104th Edition

The World's Bestselling Firearms Reference

by Jay Cassell

Published annually for more than eighty years, the *Shooter's Bible* is the most comprehensive and sought-after reference guide for new firearms and their specifications, as well as for thousands of guns that have been in production and are currently on the market. Every firearms manufacturer in the world is included in this renowned compendium. The 104th edition also contains new and existing product sections on ammunition, optics, and accessories, plus up-to-date handgun and rifle ballistic tables along with extensive charts of currently available bullets and projectiles for handloading.

With timely features on such topics as the fiftieth anniversary of the Remington Model 700, and complete with color and black-and-white photographs featuring various makes and models of firearms and equipment, the *Shooter's Bible* is an essential authority for any beginner or experienced hunter, firearm collector, or gun enthusiast.

$29.95 Paperback • ISBN 978-1-61608-874-3

ALSO AVAILABLE

Shooter's Bible Guide to Combat Handguns

by Robert A. Sadowski

For more than 100 years, *Shooter's Bible* has been the ultimate comprehensive resource for shooting enthusiasts across the board. Trusted by everyone from competitive shooters to hunters to those who keep firearms for protection, this leading series is always expanding. Here is the first edition of the *Shooter's Bible Guide to Combat Handguns*—your all-encompassing resource with up-to-date information on combat and defensive handguns, training and defensive ammunition, handgun ballistics, tactical and concealment holsters, accessories, training facilities, and more. No *Shooter's Bible* guidebook is complete without a detailed products section showcasing handguns from all across the market.

Author Robert Sadowski proves to be a masterful instructor on all aspects of handguns, providing useful information for every reader, from those with combat handgun experience in military and law enforcement fields to private citizens, first-timers, and beyond.

$19.95 Paperback • ISBN 978-1-61608-415-8

ALSO AVAILABLE

Shooter's Bible Guide to Concealed Carry

by Brad Fitzpatrick

Don't wait to be placed in a dangerous setting faced with an armed attacker. The *Shooter's Bible Guide to Concealed Carry* is an all-encompassing resource that not only offers vital gun terminology but also suggests which gun is the right fit for you and how to efficiently use the device properly, be it in public or at home. Firearm expert Brad Fitzpatrick examines how to practice, how to correct mistakes, and how to safely challenge yourself when you have achieved basic skills. Included within is a comprehensive chart describing the various calibers for concealed carry, suitable instructions for maintaining it, and, most importantly, expert step-by-step instructions for shooting.

$19.95 Paperback • ISBN 978-1-62087-580-3

ALSO AVAILABLE

Shooter's Bible Guide to Firearms Assembly, Disassembly, and Cleaning

by Robert A. Sadowski

Shooter's Bible, the most trusted source on firearms, is here to bring you a new guide with expert knowledge and advice on gun care. Double-page spreads filled with photos and illustrations provide manufacturer specifications on each featured model and guide you through disassembly and assembly for rifles, shotguns, handguns, and muzzleloaders. Step-by-step instructions for cleaning help you to care for your firearms safely. Never have a doubt about proper gun maintenance when you own the *Shooter's Bible Guide to Firearms Assembly, Disassembly, and Cleaning*, a great companion to the original Shooter's Bible.

Along with assembly, disassembly, and cleaning instructions, each featured firearm is accompanied by a brief description and list of important specs, including manufacturer, model, similar models, action, calibers/gauge, capacity, overall length, and weight. With these helpful gun maintenance tips, up-to-date specifications, detailed exploded view line drawings, and multiple photographs for each firearm, the *Shooter's Bible Guide to Firearms Assembly, Disassembly, and Cleaning* is a great resource for all firearm owners.

$19.95 Paperback • ISBN 978-1-61608-875-0

ALSO AVAILABLE

Shooter's Bible Guide to AR-15s

A Comprehensive Reference to One of America's Favorite Rifles

by Doug Howlett

There's no denying the popularity and intense fascination with AR-15s among firearms enthusiasts today. Here, inside the most comprehensive source to date, is Doug Howlett's expert approach to everything from the intriguing history of the AR to breaking down the weapon piece by piece, choosing ammunition, and even building your own gun.

In this complete book of AR-style firearms, you can peruse the products of all manufacturers, learn about the evolution of the AR from its uses in the military in the 1960s to its adaptation for law enforcement and civilian uses, and gain essential knowledge on the parts and functions of the rifle. Also included are chapters on customizing and accessorizing ARs, with special focus on small gun shops and makers and their unique and successful products. Look into the future of the AR straight from top gun authorities!

$19.95 Paperback • ISBN 978-1-61608-444-8

ALSO AVAILABLE

Shooter's Bible Guide to Cartridges

Edited by Todd Woodard

A guidebook designed specifically to teach gun users everything they need to know to select the right cartridge for their shooting needs. This title is written in an accessible and engaging style that makes research fun. The *Shooter's Bible Guide to Cartridges* is packed with full color photographs, clear and detailed diagrams, and easy-to-read charts with cartridge data.

The *Shooter's Bible* name has been known and trusted as an authority on guns and ammunition for nearly a century and has sold over seven million copies since its start. Now the *Shooter's Bible* offers readers this comprehensive and fascinating guide to cartridges. Complete with color and black and white photographs showcasing various makes and models of firearms and equipment, this guide to cartridges is the perfect addition to the bookshelf of any beginner or experienced hunter, firearm collector, or gun enthusiast. No matter what your shooting background is, you'll learn something new. This guide is a great introduction that will make readers want to seek out and get to know all the titles in the informative *Shooter's Bible* series.

$19.95 Paperback • ISBN 978-1-61608-222-2

ALSO AVAILABLE

Gun Trader's Guide, Thirty-Fourth Edition

A Comprehensive, Fully-Illustrated Guide to Modern Firearms with Current Market Values

Edited by Stephen D. Carpenteri

Gun Trader's Guide is the original reference guide for gun values. For more than half a century, this book has been the standard reference for collectors, curators, dealers, shooters, and gun enthusiasts. Now in a completely updated edition, it remains the definitive source for making informed decisions on used firearms purchases. Included are extensive listings for handguns, shotguns, and rifles from some of the most popular manufacturers, including Beretta, Browning, Colt, Remington, Winchester, and many more.

This thirty-fourth edition includes a complete index and a guide on how to properly and effectively use this book in order to find the market value for your collectible modern firearm. With new color photos from gun dealers and shows, as well as introductory materials that every gun collector and potential buyer should read, this book is the ultimate guide to purchasing firearms. No matter what kind of modern collectible firearm you own or collect, the *Gun Trader's Guide* should remain close at hand.

$29.95 Paperback • ISBN 978-1-61608-843-9

ALSO AVAILABLE

Ed McGivern's Book of Fast and Fancy Revolver Shooting

by Ed McGivern

Ed McGivern needs no introduction to gun enthusiasts and serious marksmen. For more than fifty years, he was revered as one of the top authorities in the field of small firearms. A world-champion marksmen who made *The Guinness Book of World Records*, he trained scores of law enforcement officers and developed a system of teaching that is as effective today as it was when this book was originally published. It resulted from years of experimentation and research conducted by McGivern, who utilized electric timers and other devices to determine the angles and techniques that would produce the fastest, most accurate revolver shooting. Packed with handgun lore and original photographs from the first edition, this much-sought-after classic contains a wealth of facts for marksmen everywhere.

$17.95 Paperback • ISBN 978-1-60239-086-7

ALSO AVAILABLE

Shoot

Your Guide to Shooting and Competition

by Julie Golob

Foreword by Sandy Froman

Whether you're a firearms enthusiast, an experienced shooter, or someone who has never even held a gun, *Shoot: Your Guide to Shooting and Competition* will help you explore different types of firearms, understand crucial safety rules, and learn fundamental shooting skills. This book provides an introduction to a wide variety of shooting sports through detailed descriptions that relate each type of competition to everyday activities and interests. High-quality photography from actual competitions and step-by-step instructional images augment the clearly written descriptions of both basic and advanced shooting skills.

$16.95 Paperback • ISBN 978-1-61608-698-5

ALSO AVAILABLE

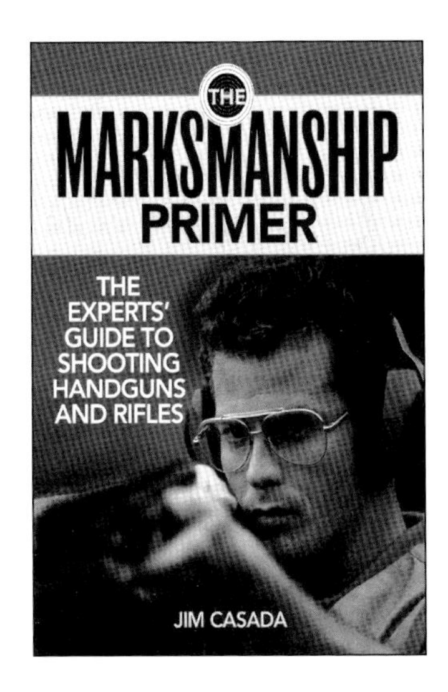

The Marksmanship Primer

The Experts' Guide to Shooting Handguns and Rifles

by Jim Casada

The Marksmanship Primer serves as a road map to greater shooting proficiency as well as greater enjoyment of the sport of shooting. Jim Casada, renowned outdoors author and editor, has brought together the best selections from America's great gun writers of yesterday and today. Marksmen of all levels of experience—beginners, pros, and hobbyists—can benefit from this collection of shooting wisdom.

Topics include:

- Positions for Rifle and Handgun Shooting
- Sighting In
- Ballistics
- Rifle Marksmanship for the Hunter
- Accuracy at All Distances
- Hunting with the Handgun
- Physical and Mental Fitness for the Marksman

$14.95 Paperback • ISBN 978-1-62087-367-0

ALSO AVAILABLE

Sixguns and Bullseyes and Automatic Pistol Marksmanship

A Comprehensive Manual on Target Shooting

by William Reichenbach

Whether you're a target shooting enthusiast, an experienced shooter, or someone who has never held a gun, *Sixguns and Bullseyes and Automatic Pistol Marksmanship* will help you explore different types of handguns, fundamental shooting skills, and expert tips to gain marksmanship precision.

This edition combines two classic shooting manuals from the 1930s in one volume for modern audiences. Author and gun enthusiast William Reichenbach's conversational, down-to-earth writing style makes this primer very approachable to all types of readers and shooters. He describes his seven key points—hold, stance, relaxation, moving the gun into position, sighting, squeeze, and breathing—as a basis to target shooting, as well as other topics, including:

- Ascent to the Olymp
- Time and Rapid Fire
- Trimming Your Gun
- Ammunition Wrinkles
- The Ideal Automatic
- The "Draw"
- Preparing for the Fray
- Homo Sapiens and Other Game

$14.95 Paperback • ISBN 978-1-62087-372-4

ALSO AVAILABLE

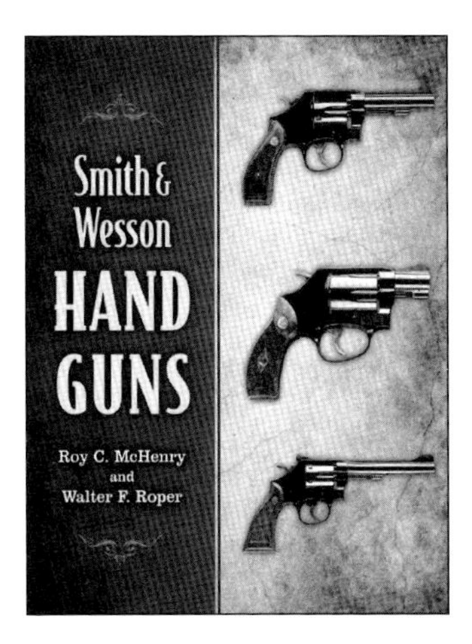

Smith & Wesson Hand Guns

by Roy C. McHenry and Walter F. Roper

"The story of Smith & Wesson handguns and their evolution is one of the hallowed tales of American firearms' history," according to the firearms writer Jim Casada. Anyone who collects Smith & Wessons or is simply interested in their backstory will cherish this book.

Though originally published in 1945, more than half a century ago, *Smith & Wesson Hand Guns* remains the source for Smith & Wesson enthusiasts. It is an authoritative reference and has remained, for over five decades, the cornerstone upon which Smith & Wesson research rests. This work is foundational, supported by sixty-three detailed illustrations showing the handguns, the unique hammer mechanism, and facsimile reproductions of vintage advertising copy.

The first twenty-four chapters of the book, which tell the story of Smith and Wesson and the development of Smith & Wesson handguns, are very informative. After the reader becomes familiar with Smith and Wesson's history together, as well as their creation of a business, illustrations exhibiting Smith & Wesson handguns will show rather than tell of their magnificence. Finally, descriptions of different caliber guns are given, where readers will gain invaluable information regarding Smith & Wesson handguns. For any Smith & Wesson enthusiast or collector, this work is impossible to put down.

$12.95 Paperback • ISBN 978-1-62087-715-9

ALSO AVAILABLE

Taking Your First Shot

A Woman's Introduction to Defensive Shooting and Personal Safety

by Lynne Finch

Numbers don't lie; more and more women are purchasing guns and learning to shoot! While shooting used to be a male-dominated sport, women across the country have begun discovering that a trip to the range not only is relaxing, but also brings with it a sense of strength and empowerment. *Taking Your First Shot* is an introductory guide perfect for either those stepping out onto the range for the first time or those looking to brush up on their skills. Author Lynne Finch coaches women on the decision to learn to shoot, how to find formal training, selecting and purchasing a handgun, defensive versus practice ammunition, storing and caring for your gun, and concealed carry options.

Along with learning the shooting basics, Finch also teaches readers the importance of situational awareness and the basics of self-defense. Sometimes a gun isn't always an answer, and it's important to have a proportional response to the situation. Finch begins with teaching readers how to become aware of their surroundings, what to watch for, and how to respond. From there, she goes on to define proportional response and why carrying pepper spray, a kubotan, or even a whistle can make all the difference.

Learning to shoot is a personal decision, but with the proper training and practice, shooting can become both an enjoyable and liberating sport.

$14.95 Paperback • ISBN 978-1-62087-717-3

ALSO AVAILABLE

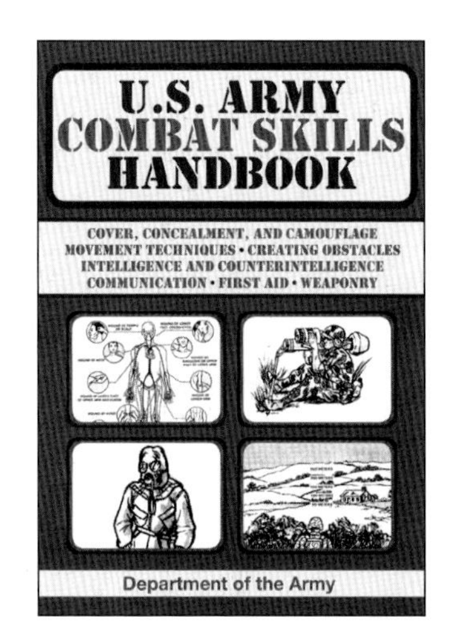

U.S. Army Combat Skills Handbook

by Department of the Army

Recognizing that "wars are not won by machines and weapons but by the soldiers who use them," this comprehensive manual not only informs the reader of the timeless skills necessary to survive on the battlefield, but also instructs the soldier on how to perform and execute these tasks to succeed in combat. From concealment and mobility to first aid and personal care, you too can possess the knowledge armed service people are equipped with so they can do their jobs properly and confidently.

This informative and comprehensive guide draws upon the real-life experiences of soldiers who have faced warfare and lived in combat zones.

Practical and explicit instructions on team formation, proper positioning for offensive and defensive maneuvers, and handling of equipment and weaponry are thoroughly explained. Also included is advice on life-saving CPR and wound-care techniques; sections on combat intelligence and nuclear, biological, and chemical warfare; and appendices covering mines, demolitions, obstacles, combat in urban areas, tracking, and evasion and escape. Illustrations throughout the book depict various field scenarios that soldiers face in a war zone. See what it takes to perform, inspire, and lead in the U.S. Army!